Principles of Modern Physics

Kunming Gu & Tianze Gu

If you love science,

please take this book home.

If your ambition is to contribute to mankind,

please take this book home.

If you also want to win the Nobel Prize,

please take this book home.

Maybe you found this book by accident,

Please also take her home.

Maybe she will change your life track,

Open a door, pave a way and light a lamp for you!

ISBN: 978-1-957144-77-1

First paperback edition March 2023.
Printed in the United Stated of America.

Front cover image by Li Hui
Book Layout design by Li Hui

Published by Asian Culture Press LLC
1942 Broadway, Suite 314C
Boulder, CO 80302
United States

CONTENT VALIDITY:

Modern physical theory is not a complete theory, especially the overall self-consistency. The incompleteness of the theory is not limited to the quantum mechanics that describes the microstructure, but also includes the macroscopic theory based on "relativity". For example, light theory, space theory, etc.

In the theory of relativity, "the speed of light is constant" is put forward as a basic assumption. In fact, the "constant speed of light" is due to the Doppler redshift of light. The reason for the Doppler redshift of light is that "the speed of light does not change". This article will take you to understand the physical principle of this mechanism.

Through cosmological redshift, we can draw the conclusion that vacuum space is also a form of the existence of matter. This paper describes the physical characteristics of space material form from different physical angles.

This book will also tell you why matter has mass, how inertia is formed, and why "gravitational mass" is equal to "inertial mass". In addition, all doubts in modern physics can be answered in the book, including "dark energy", "dark matter", quantum entanglement, etc.

PREFACE

Modern physical theory is a physical theory based on "relativity" and "quantum mechanics". Einstein's "theory of relativity" describes the movement of the world from a macro perspective, while "quantum mechanics" describes the structure of matter from a micro perspective. Modern physics focuses on describing the world from both macro and micro aspects.

Originally, modern physical theory should be in the ascendant in the development of modern society, but the fact is that the progress of modern physical theory in all aspects is not good. In addition to being unable to explain some physical phenomena (such as quantum entanglement), the theory also seems to have the problem of self-consistency. Today, after a theoretical system supporting modern science and technology has developed for more than 100 years, we have to re-examine the completeness of the theoretical system. If a theoretical system is not self-consistent, it often needs ten times more reasons to justify itself. The imperfection of the theoretical system will lead to the obscurity of modern physical theory. Therefore, Feynman said, "no one really understands quantum mechanics".

We still remember the two dark clouds in the sky, "blackbody radiation" and the Michelson Morey experiment, which are the origin of modern physical theory. The present problem is related to the original concept of light. Light not only reflects the secret of microstructure, but also involves the physical essence of vacuum space. Light and space are the basis for human beings to understand nature.

Through the study of the Doppler red shift, cosmological red shift and the principle of "light speed unchanged", it is found that the original understanding of light is biased. And this deviation leads to the wrong cognition of space. From the relationship between the Doppler red shift of light, the cosmological red shift and the "constant speed of light", we can get a new concept about space: vacuum space is not a hollow, but a physical entity.

Newton explained the phenomena of celestial bodies and oceans by the action of gravity, but he did not explain the cause of gravity. Maxwell equation gives the relationship between electric field and magnetic field, but electric field and magnetic field are not independent substances. Einstein's field equation describes the space under the action of gravity. If the space is a hole, how to produce physical effects. When we regard vacuum space as a physical entity, all problems will be solved.

According to the propagation characteristics of light and space, a new spatial structure model can be built. Using this space model, we can find out the cause of gravity. Answer why all substances have "quality"; what is inertial mass; what is the physical mechanism of "inertial mass"; what are "dark energy" and "dark matter"; why there is entanglement between quanta. Under this space model, all physics puzzles can be solved.

"Principles of modern physics" tells two principles. One is the principle of physics itself, and the other is the physical principle of world movement. This paper analyzes three forms of matter in the universe: equivalent equilibrium state, non-equilibrium state and absolute equilibrium state. The three forms of matter describe the three different "physical entities" of the sun, moon, stars, black holes and vacuum space.

"Equivalent equilibrium state" can well describe the movement state of microstructure. Why does microstructure have wave particle duality? What is the quantum superposition state? What problem does the uncertainty principle reflect.

The "non-equilibrium state" introduces the evolution of matter and how matter eventually forms a black hole. It perfectly interprets some special astronomical phenomena such as supernova explosions and Quasar jets.

The "absolute equilibrium state" unravels how vacuum space, like "God", dominates and regulates the movement of matter and forms various physical laws. The article not only explains the structure of the universe, but also describes the physical laws of its operation.

"Principles of modern physics" is a breakthrough in modern physical theory. This theoretical progress will promote the development of the whole science and technology. In addition to the technical application, the principle can also provide theoretical support for "cosmic observation", "black hole" research, "dark matter" detection and so on.

Nature always creates the world in the simplest way. The principles described in the book "principles of modern physics" can be understood as long as you have certain physical knowledge and understand a little higher mathematics. It is not only the knowledge that professional scientific and technological personnel should master, but also the "knowledge structure" that everyone should have. It will tell you what is going on in the world around us. Learning this science will make you smart.

CONTENTS

PART ONE

Doppler redshift and the principle of "constant speed of light"

Humans understand the objective world around them mainly through light. Light not only transmits the "shape" and "color" of the world, but also reflects the physical essence of the world. We need not only to understand the world through light, but also to understand why light can reflect the objective world. What is the physical nature of light?

Light, opened the chapter of modern physics and laid the foundation of modern physics. In the development of modern physics, the theory of light has always been controversial. The current theory of light is also incomplete, mainly including the principle of "constant speed of light", wave particle duality of light, etc. In fact, until today, we have not really understood the nature of light and why the speed of light remains unchanged. Light is everywhere, but what is it? Today, let's review the controversy about light theory in history again, and explore the secrets of light again.

§ 1·1 Reference frame and electromagnetic space

If we need to describe the motion of a particle, we must select a motion reference. The description of the motion of the same particle will be different if the reference object selected is different. In our life, we can choose the ground as the reference object, or we can choose a moving object to describe the relative motion of another object. But this is to describe the motion of objects under the premise of the Earth reference system.

When leaving the earth reference system, everything is moving, the sky is moving and the earth is moving. How can we describe the motion track of a star in space? Can the motion of microscopic particles also be described in the Earth reference system? Does the speed of light have a moving reference?

How to define the frame of reference of motion was a philosophical proposition in the 19th century. Newton believed that the reference object of motion should be an absolute still space. The most famous is his bucket experiment. Before the Michelson Morey experiment, most physicists believed that the frame of reference for motion was ether. Ether is absolute still space.

Mach believes that motion is relative. You sit at home, you are moving relative to the sun, you are not moving relative to the earth. Everything in the universe is in motion, and there is no absolutely static motion reference frame.

Galileo scientifically summarized the relativity of motion. The principle of "relativity" is put forward: the mechanical law is equivalent in all "inertial coordinate systems". The mechanical process is exactly the same for the static "inertial system" and the moving "inertial system".

In the above point of view, the movement identified by Newton is the movement with energy attribute, and the movement is from never moving to moving. The movement from never moving to moving is an absolute movement, which symbolizes energy. Mach's definition of motion is the motion we need to describe, which is specific. When we study the motion of matter, we must choose the reference object of motion. Therefore, the motion of selecting the reference object is a relative motion. Galileo's "relativity" principle defines this movement from a scientific perspective.

For the "inertial system", whether the "inertial system" is stationary or moving at a uniform speed of 30 kilometers per second, the physical laws in the "inertial system" are the same, and the mathematical form describing the physical laws is the same. When you stand on a train moving at a constant speed and throw a ball, the trajectory of the ball relative to the train is the same as that of the ball relative to the ground when you stand on the ground and throw a ball.

Because the laws of motion of objects are the same in any "inertial frame". We do not need to consider whether the "inertial system" is in a stationary state or in uniform motion. We can use the same mathematical formula to describe the trajectory of an object. This is the first hypothesis of Einstein's "special relativity", that is, the principle of relativity.

For any "inertial system", we can use the same mathematical formula to describe the trajectory of object motion. If this reference system is a non "inertial system", this rule cannot be established. Another problem is how to carry out "coordinate transformation" between two "inertial system" with different motion speeds. Therefore, Einstein introduced a second hypothesis, "the speed of light is constant".

Including Einstein himself, the general view is that the space-time view of "special relativity" is a negation of Newton's absolute space-time view. In fact, this view is incorrect. Einstein's space-time view of "special relativity" not only did not deny Newton's absolute space-time view, but also developed Newton's physical concept of absolute stationary space and enriched its physical connotation. Newton's idea of absolute stationary space is "constant speed of light" in "special relativity".

Why is it that the assumption of "the speed of light is constant" in Special Theory of Relativity is Newton's absolute stationary space reference system? Let's start with the connotation of "constant speed of light". The principle of "constant speed of light" means that the speed of light in vacuum is the same for any observer. In any reference frame, the speed of light in vacuum is constant. Measured in different "inertial frames", the speed of light is the same value. In this way, light becomes the common reference speed of motion between different "inertial frames". Light plays the role of "reference" or "standard ruler" here.

Why should we stipulate that the speed of light is the same value in any reference system? Only by specifying that the speed of light is the same value in any reference system can the relationship between each motion reference system be established. Transformation factors between different reference systems γ Is:

$$\gamma = \frac{1}{\sqrt{1-\frac{v^2}{c^2}}} \qquad 1-1$$

It can be seen from formula 1-1 that if the motion speed of the reference system is v, the speed of light is C, the v ratio to C is the basis of Lorentz transformation. The speed of light C plays the role of international currency or standard ruler in Lorentz transformation.

The transformation between different motion speed reference systems is similar to international trade. When trade is settled in that currency, it means that the basis behind trade is gold or oil. Since the speed of light plays the role of international currency or standard ruler in Lorentz transformation, the speed of light is the speed of light propagation in vacuum space, so vacuum space is the background space of Lorentz transformation.

In order to distinguish the concepts of different spaces and make space have a strict physical definition, we can define the space carrying and transmitting electromagnetic waves in "vacuum space" as: electromagnetic space. Thus, "electromagnetic space" is the background space for Lorentz transformation in Special Theory of Relativity.

The background space of Special Theory of Relativity is Minkowski Space time. When the speed of light is used as the basis for Lorentz transformation, "electromagnetic space" replaces Minkowski Space time as the background space of Special Theory of Relativity. The "electromagnetic space" not only meets the spatial characteristics of Minkowski Space time, but also has the spatial characteristics of Newton's "absolute static space". Why "electromagnetic space" has these spatial characteristics will be specifically discussed later.

When recognizing that the speed of light is a constant value in different "inertial systems" and conducting Lorentz transformation, we have unintentionally taken the "electromagnetic space" as a reference for motion. When "electromagnetic space" is used as a reference for motion, it actually unifies the views of Mach and Newton.

Mach's emphasis on motion is "relative motion", and "relative motion" emphasizes the relativity of motion. The motion defined by Newton is "absolute motion", which reflects the energy attribute of motion. The two assumptions in Special Theory of Relativity represent Mach's "relative motion" and Newton's "absolute motion" respectively. Transformation factors in special relativity γ, It not only reflects the relativity of motion, but also contains the energy attribute of motion.

The two assumptions in Special Theory of Relativity are well integrated. "Relativity" principle describes the relativity of motion and also defines the reference object of motion. For example, the traveling speed of a train is the speed of motion relative to the ground. The reference object of motion can only be the ground, not the train station. The reference object of passenger movement in the train can only be the train, not the passenger's home. When light travels in vacuum, the moving reference object can only be "electromagnetic space", not the light source or observer. When the light we received was emitted by celestial bodies 5 billion years ago, the Earth was not yet born. The speed of light cannot take the earth as a reference for its motion. When light travels in vacuum, the "relativity of motion" determines that vacuum space is the moving reference of light.

The relativity of motion gives the possibility of coordinate transformation between different "inertial systems". The principle of "constant speed of light" actually changes the background space of coordinate transformation of different "inertial systems" into "electromagnetic space". Without considering these factors, "the speed of light is constant" actually includes three aspects: 1. The speed of light propagation in vacuum is a constant. 2. The speed of light emitted when the light source is stationary is the same as that emitted when the light source is moving. 3. The speed of light measured when the observer is stationary is the same as that measured when the observer is moving.

he above three items reflect the three elements of the principle of "constant speed of light": moving reference, light source and observer. The relative physical quantity between the light source and the observer is related to the physical behavior of light, and the relative speed between them must be based on the physical relationship determined by the speed of light. There is no direct correlation between the light source and the observer, and there is no relativity.

The motions we describe in the "inertial reference system" are relative motions. The speed of motion is the relative speed of motion. Light represents a kind of energy, and "speed of light" is a kind of "absolute motion" speed. The moving reference of the speed of light must be Newton's absolute space of rest. After excluding the existence of the ether, the absolute still space is vacuum space or "electromagnetic space".

The reference object of light speed is "electromagnetic space", and the moving reference object of light source and observer must also be "electromagnetic space". "Electromagnetic space" is the background space of all movements and also the reference object of all movements. This relationship not only contains Mach's relative principle, but also contains the physical connotation of Newton's absolute space-time concept. The motion described in this frame of reference has not only relativity, but also energy attribute.

When the "electromagnetic space" is taken as the moving reference of light, the relative motion of the light source becomes the relative motion of the light source and the "electromagnetic space". The relative motion of the observer becomes the relative motion between the observer and the "electromagnetic space". The relative motion between the light source and the observer is equal to the superposition of the relative motion velocity of the light source, the observer and the "electromagnetic space".

When we agreed to measure in any "inertial frame" and the speed of light in vacuum is the same constant C, we inadvertently confirmed that the reference frame of motion of light speed is "electromagnetic space". If there is no "electromagnetic space", v in Formula 1-1 cannot be defined. Only after the "electromagnetic space" is set as the moving reference object, can it be determined v Size of. That is v It is the same moving reference object as C, both of which are electromagnetic spaces. With the moving reference system of "electromagnetic space", the principle of "constant speed of light" includes three aspects:

1. Light propagates in "electromagnetic space", and the propagation speed is a constant C relative to "electromagnetic space".

2. No matter whether the light source is in a static state or a moving state relative to the "electromagnetic space", the propagation speed of the electromagnetic wave radiated by the light source to the "electromagnetic space" is the same constant C.

3. Whether the observer is stationary or moving relative to the "electromagnetic space", the speed of light received by the observer from the "electromagnetic space" is the same constant C.

With "electromagnetic space" instead of "absolute static space" as the motion reference of "absolute motion", this not only defines the speed of light, but also defines the motion properties of celestial bodies. We all know that space is expanding. Because electromagnetic space and vacuum space are the same subject, when vacuum space expands, electromagnetic space also expands. Therefore, when we take electromagnetic space as the reference of motion, motion has a clear physical definition. When a moving particle is stationary relative to the "electromagnetic space", the particle is actually in an "absolute stationary" state. When a particle moves relative to the "electromagnetic space", the particle is in the "absolute motion" state. If the motion reference system selects other motion reference systems, the motion of particles relative to other motion reference systems is called "relative motion". Relative motion may include absolute motion, but absolute motion is only for relative motion in electromagnetic space.

Obviously, "relative motion" is the motion defined by Mach, and "absolute motion" is the motion defined by Newton. It is very useful in the study of "astrophysics" to distinguish two kinds of motions with different properties. For example, there is "relative motion" between Galaxy A and Galaxy B, but there may be "absolute motion" between Galaxy A and "electromagnetic space", or it may be in a state of synchronous expansion with "electromagnetic space". Although Galaxy A has "relative motion" against Galaxy B, Galaxy A is actually in an "absolute static" state. This involves the relationship between "energy" and "dark energy" in the universe.

As a motion reference frame, electromagnetic space has a very profound physical connotation, which determines the background space of the physical system we describe. For the propagation of light, the "electromagnetic space" reference system defines the "speed of light", and for the motion of celestial bodies, the "electromagnetic space" reference system defines the nature of celestial body motion. The reference frame of "electromagnetic space" lays the foundation of the whole modern physical theory.

§ 1·2 Physical principle of Doppler redshift

Because the Doppler redshift is similar to the Doppler effect, the Doppler frequency shift formula of sound waves is often used to calculate the Doppler redshift of light when analyzing the Doppler redshift, but this is not correct. The Doppler frequency shift of sound waves is caused by the relative motion between the wave source and the receiver. The principle of "constant speed of light" tells us that no matter whether the light source and the observer move away from each other or close to each other, the speed of light propagation between the light source and the observer is C. The movement of light source and observer away from each other or close to each other will not change the speed of light propagation. The velocity transformation formula of acoustic Doppler effect is not applicable to the red shift of light spectral lines.

Some people also use "special relativity" to explain the red shift of the spectrum or the principle of "constant speed of light". The principle of "constant speed of light" is the basic condition set by "special relativity". Lorentz transformation factor is derived from the principle of "constant speed of light". We can neither prove the "constant speed of light" with the derived results, nor calculate the red shift of the spectrum with Lorentz transform.

The correct understanding of the Doppler redshift of light lies in the true understanding of the nature of light. The Doppler effect of light is only related to the moment when the light source radiates electromagnetic waves into space, or the moment when the observer receives electromagnetic waves from space. At this time: if the light source moves relative to the "electromagnetic space", the light emitted by the light source into space will have a red shift or a blue shift; when the observer moves relative to the "electromagnetic space", the light received by the observer from the space will also have a red shift or a blue shift. Obviously, the reference object of relative motion here is "electromagnetic space".

The essence of object luminescence is that the light source radiates and releases energy to its surrounding space, that is, energy transfer. The Doppler redshift of light is: if the illuminant moves relative to the "electromagnetic space" around it, the wavelength of light changes when the light radiated by the illuminant enters the "electromagnetic space" in the form of electromagnetic wave. See Figure 1.1:

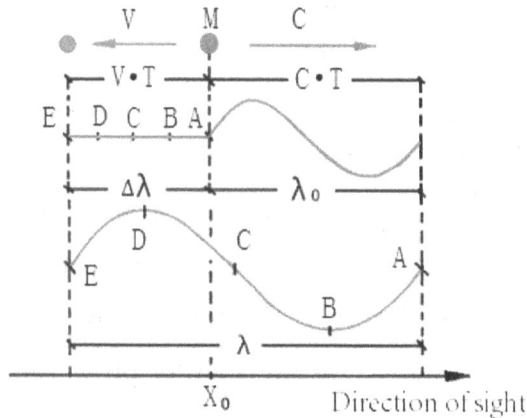

Fig.1.1

In Figure 1.1, M: light source V: light source motion speed C: light speed T: period.

Let X_0 be a point in space-time. A light source M is stationary at X_0. When time $\Delta t=T$, the light wave length radiated by the light source is λ_0. If the light source moves in the opposite direction of the viewing direction at V speed. In one cycle, the light source will move from point A, B, C, D to point E.

The length of A-E segment is the space distance that the light source moves reversely in a cycle.

Due to the principle of constant speed of light, the wavelength of radiation after the motion of light source M is still λ_0. Points A, B, C, D and E on the motion track of light source M correspond to points A, B, C, D and E on the radiation waveform respectively (see Figure 1.1).

In Fig. 1.1, although the electromagnetic wave length radiated by the moving light source is still λ_0, the radiation wave formed in the observation direction space becomes λ due to the relative motion between the light source and its electromagnetic space. The interrelationships are:

$$\lambda = \Delta\lambda + \lambda_0 \qquad 1\text{—}2$$

$$\frac{\Delta\lambda}{\lambda_0} = \frac{V}{C} \qquad 1\text{—}3$$

$$Z_D = \frac{\Delta\lambda}{\lambda_0} \qquad 1\text{—}4$$

λ_0: Original Wavelength of light wave $\Delta\lambda$: Wavelength change λ: Wavelength after red shift Z_D: Doppler redshift

It can be seen from the above that when the light source moves relative to its surrounding electromagnetic space, the wavelength of the electromagnetic wave radiated by the moving light source changes at the moment of entering the electromagnetic space. From λ_0 to λ, the difference between them is $\Delta\lambda$. $\Delta\lambda$ is equal to the space distance that the light source moves in a period. Namely:

$$\Delta\lambda = T*V \qquad 1\text{—}5$$

V in formulas 1-3 and 1-5 is the relative motion speed between the light source and the "electromagnetic space", but in the eyes of the observer, this relative speed is the motion speed between the light source and the observer. Therefore, if the moving direction is not on the same line with the observation direction, the moving speed V should be taken as "radial speed", that is, the speed component of the object moving speed in the direction of the observation line of sight.

When the observer receives the electromagnetic wave of light from the "electromagnetic space", if there is a relative movement between him and the electromagnetic space, this relative movement will also produce a red shift. The physical mechanism of red shift during reception is the same as that when the light source emits light. When the light source emits light, the light source moves along the direction of light propagation, resulting in a blue shift. When the observer receives light, if the observer moves along the direction of light propagation, it will produce red shift. Whether it is a light source or an observer, if the moving direction and the light propagation direction are not in the same line, the relative moving speed should be converted into "radial speed".

In fact, the Doppler redshift observed by the observer is the redshift jointly generated by the light source and the observer when transmitting and receiving respectively. The red shift value of its spectrum directly reflects the total "radial velocity" between the light source and the observer. The v in Formula 1-3 and Formula 1-5 is the total "radial velocity" in practical application. The total "radial velocity" is the superposition of the velocity values after the relative motion velocities of the light source and the "electromagnetic space" and the observer and the "electromagnetic space" are converted into "radial velocity".

§1·3　Principle of constancy of light velocity

Our eyes can see a complete world, not only because the world is full of sunshine, but also because "the speed of light remains unchanged". If the speed of light obeys the principle of speed superposition, the world we see will be a messy image. The image of moving objects will be misplaced. Fast moving objects enter our sight first, and slow—moving objects enter our eyes later. We will not be able to see a complete picture of the world in motion. Due to nuclear fusion, the sun is no longer a radiant disk. What we see first will be the solar flare. We can see that there is a complete world around us, all proving the principle of "constant speed of light".

Why does the speed of light remain the same? Although there are various theories, we can't really understand the principle of "constant speed of light" without really mastering the Doppler redshift mechanism of light. The principle of "constant speed of light" and Doppler redshift are the positive and negative sides of a coin. One side is represented by Doppler redshift, and the other side is "constant speed of light".

Figure1.1 not only shows the physical mechanism of Doppler redshift, but also reveals the physical principle of "constant speed of light". When the relative motion of the light source or the observer causes the light to produce a Doppler red shift, the speed of light will remain unchanged.

The principle of "constant speed of light" includes three aspects·

1. Light travels in electromagnetic space, and its velocity is a constant C relative to electromagnetic space. "Electromagnetic space" is the carrier and motion reference frame of electromagnetic wave. The propagation speed of light in "electromagnetic space" is determined by the spatial dielectric constant ε_0 and spatial permeability μ_0 decision. Dielectric constant ε_0 and spatial permeability μ_0 is a physical property

of vacuum space. It not only shows that space has physical properties, but also shows that light propagates in vacuum space, and its speed has nothing to do with the movement of light source and observer.

2. Whether the light source is stationary or moving relative to the electromagnetic space, the propagation speed of the electromagnetic wave radiated by the light source to the electromagnetic space is the same constant C. It can be seen in Fig. 1.1 that when the light source M is stationary at X_0, the speed of the light emitted by it in the "electromagnetic space" is C; when the light source M moves at the speed v, the speed of the emitted light is still C after entering the "electromagnetic space". Whether the light source M is in a stationary or moving state, the speed of its electromagnetic wave propagation in the "electromagnetic space" is C. The motion of light source M only changes its wavelength, and does not affect the speed of light.

3. Whether the observer is stationary or in relative motion relative to the electromagnetic space, the speed of light received by the observer from the electromagnetic space is the same constant C. When there is relative motion between the observer and the electromagnetic space, this motion will only change the wavelength of light without affecting the propagation speed of light.

The principle of "constant speed of light" is composed of 1, 2 and 3 basic principles. The principle of "constant speed of light" and Doppler redshift are two aspects of the same mechanism. This mechanism reflects the relationship between light source, observer, light wavelength and light speed. Because the speed of light in the "electromagnetic space" is constant, it causes the red shift of the spectrum. On the contrary, because the light source or the reference system of the observer moves relative to the "electromagnetic space", the light has a red shift, so the "speed of light is unchanged".

According to the above principle, we will do a thought experiment: if we have a spaceship that can fly at the speed of light, if we send a beam of light on the spaceship, what will happen to this beam of light?

We should first make clear two concepts, geometric space and "electromagnetic space". Geometric space refers to the geometric characteristics of vacuum space, which reflects the space distance. "Electromagnetic space" is the space where electromagnetic waves propagate, which reflects the electromagnetic characteristics of vacuum space. The spaceship flies in the vacuum space at the speed of light, passing the geometric distance of space. The propagation path of light is "electromagnetic space". Although the spacecraft flies at the speed of light in geometric space, this speed will not be superimposed on the speed of light propagation in "electromagnetic space". The speed of light propagation in "electromagnetic space" is still C.

The speed of the spacecraft in geometric space will not be superimposed on the speed of light propagation, but will change the wavelength of light. If the spacecraft moves in the same direction as the light, the light will produce a blue shift. The blue shift of light is the increase of frequency. It can be seen from $E=h·v$ that the energy possessed by the flight speed of the spacecraft will be superimposed on the energy of "photons". If the speed of the spaceship reaches the speed of light, the blue shift of light will reach "- 1". At this time, the wavelength of light tends to zero, the frequency will increase infinitely, and the energy of "photons" will also increase infinitely.

What if the spaceship flying at the speed of light receives the light traveling in the same direction? The spaceship flies in the same direction as the "photon". If the spaceship reaches the speed of light, the spaceship in the geometric space will go hand in hand with the light in the "electromagnetic space". The speed of the light received by the spaceship and the propagation of this light relative to the "electromagnetic space" is still C. But the light received by the spacecraft will have a red shift. When the spaceship reaches the speed of light, the light received by the spaceship, its wavelength will tend to be infinite, and the energy of light will tend to zero.

In the above theory, "electromagnetic space" is an important concept, which involves light, and the reference frame of relative motion related to light and. Without such a motion reference system, we cannot define motion. "Electromagnetic space" is related to light, speed of light, redshift, "constant speed of light" and other physical quantities. Through "electromagnetic space", we can obtain the physical laws that can summarize the two physical mechanisms of Doppler redshift and "constant speed of light". "Electromagnetic space" not only determines the physical characteristics of light, but also determines the motion characteristics of light. If the physical quantity "electromagnetic space" is lost, the above laws cannot be established.

The reason why "electromagnetic space" is light and the motion reference system of relative motion related to light is that it has the spatial nature of Newton's absolute static space and has the physical essence similar to "ether". The light we describe is the relative motion of light. We need to select a reference object. But the essence of light is energy, and the motion of light is "absolute motion". Therefore, its reference must be an absolute static space similar to Newton's.

It is necessary to mention the "ether" again. Before Michelson Morey's experiment, the "ether theory" was a mainstream view recognized by most scientists. After Michelson Morey's experiment, the "ether theory" was completely denied. This result not only covers up the physical essence of light and space, but also blurs the physical concept of "constant speed of light". Because of this, Michelson Morey experiment changed the research direction of physics and affected the development of physics, especially quantum mechanics.

For the Michelson Morey experiment, some people have put forward different voices, but they have not been able to overturn this experimental result in theory or experiment. Although Einstein created the theory of relativity because of this experiment, the principle of "constant speed of light" has been misinterpreted.

Does the transmission of light really need no media? Can the results of Michelson Morey experiment prove that "Ether" does not exist? Now we use the physical principle of Doppler redshift of light to analyze the Michelson Morey experiment.

§ 1·4 Michelson —Morey experiment

Like the interference and diffraction of light, the Doppler redshift of light is also a common physical phenomenon in nature. As long as there is relative motion between the light source or observer and the motion reference frame of light, in order to ensure that the speed of light remains unchanged, the light will produce red shift or blue shift. Since the Doppler redshift of light is a common phenomenon, will it affect the experimental results in the Michelson Morey experiment? Let's analyze it in detail.

Michelson Morey experiment is a physical experiment conducted by Albert Michelson and Edward Morey from 1881 to 1884 to measure the relative velocity of the earth and the ether. Figure 1.2 shows the experimental principle of the device.

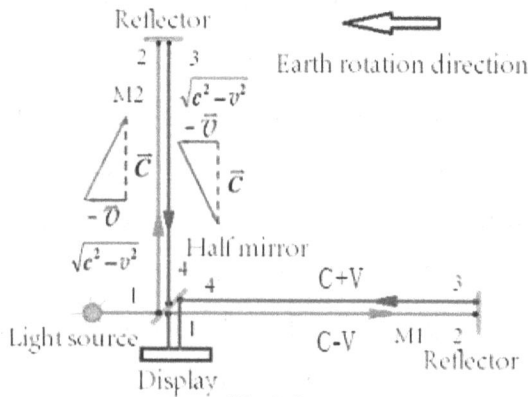

Fig .1.2

After passing through the spectroscope, the light source is divided into M1 and M2, which are perpendicular to each other. In M1 optical paths 1 to 2, light moves in reverse with ether, so the signal transmission speed is C-V. In 3 to 4, the direction of light is the same as that of ether, and the transmission speed of optical signal is C+V. Therefore, in the whole optical path of M1, the influence of ether on light propagation speed cancels each other.

In the M2 optical path, the optical signal transmission speed between 1 and 2 is $(c^2-v^2)\wedge1/2$, and the signal transmission speed between 3 and 4 is also $(c^2-v^2) \wedge 1/2$, where V is the movement speed of the reflector relative to the ether. Between 1, 2 and 3 and 4, the influence of ether on the speed of light is "-V". Therefore, when the light distances of M2 and M1 are equal, the signal will have a delay when the M2 light path reaches the observation screen. When the two beams reach the observation screen, there will be phase difference, and the phase difference will change with the rotation of the experimental device. The interference fringes formed on the observation screen will also change with the rotation of the experimental device. However, no change of interference fringes was observed in this experiment. Therefore, the experimental results deny the existence of ether. (Please refer to Michelson-Morley Experiment)

If the physical concept of "electromagnetic space" is used to replace ether, they should have the same physical effect. Now we analyze whether the experimental device can obtain the expected experimental effect according to the physical principle of Doppler red shift.

In the Michelson Morey experimental device, we need to pay attention to the physical mechanism of light at the spectroscope and reflector. When light passes through a spectroscope or reflector, it passes through a physical interface. On one side of the physical interface, light is an electromagnetic wave in vacuum, and on the other side of the physical interface, light is an energy particle that can be refracted or reflected. When the light acts on the mirror, there is a wave particle conversion process. The conversion process of light is similar: the observer receives light and the light source emits light. In these two action instants, the light may have a red shift or a blue shift.

Light in vacuum is an electromagnetic wave or an energy package. When light shines into the mirror, it is to transfer this energy of light to the mirror. The refraction or reflection of light is that the mirror will receive energy and re radiate it into space. Therefore, the process of light refraction and reflection includes two behaviors of light reception and radiation. In the process of receiving and radiation, if the mirror moves relative to the electromagnetic space, the refraction or reflection of light will produce a red shift or a blue shift during the energy transfer process of receiving and radiation.

The direction vectors of the two paths of light in Michelson Morey experiment are perpendicular to each other. Due to the earth's rotation or other relative motion, if the direction vector of one path of light is parallel to the direction of electromagnetic space motion, the direction vector of the other path of light must be perpendicular to the direction of electromagnetic space motion.

According to Huygens principle, the refraction interface and reflection interface can be divided into 1, 2, 3 and 4 interface points for analysis. In Figure 1.2, points 1 and 3 can be regarded as a new point light source, and points 2 and 4 are the receiving points of light.

In Figure 1.2, the spectroscope is the origin and convergence point of the two beams of light. Point 1 in the optical path M1 can be regarded as a point light source. With the rotation of the earth, the experimental device and spectroscope can be regarded as the movement of the light source relative to the electromagnetic space. If the light source moves to the right, the electromagnetic space moves to the left. In this way, at point 1 of M1 light path, the light source moves along the direction of light transmission. According to the principle of Doppler red shift, the blue shift will occur when the light source moves along the direction of light transmission.

Point 2 of optical path M1 is the light receiving point. When the light receiving point moves along the light propagation direction, it will produce a red shift.

According to the previous redshift formula, the V that determines the redshift and the V that determines the blue shift are the moving speed of the earth relative to the electromagnetic space. Therefore, the resulting blue shift and red shift just offset each other. If point 2 in the M1 light path is a human eye, he can neither see the blue shift nor the red shift of the light. He did not know that light had undergone two wavelength changes. The light he saw was still the original wavelength. In fact, the light he saw has undergone a blue—shift and a red—shift.

Points 3 and 4 in the optical path M1 are the same as points 1 and 2. Point 3 is the luminous light source, which moves against the light direction to produce a red shift. Point 4 is the light receiving point. Moving towards the light direction will produce a blue shift. Therefore, in the M1 optical path, the light goes out from point 1 and comes back from point 4, and the wavelength and phase do not change.

Figure 1.3 a show that due to the relative motion between the earth and electromagnetic space, the M2 light path is a slash. Since the optical path of M2 is greater than M1, the M2 optical signal will have a delay from the origin to the convergence point relative to M1, resulting in a phase difference with M1.

Fig. 1.3b is the direction vector between points 1 and 2 of the

M2 optical path, which can decompose the parallel and vertical components. The parallel component represents the direction vector of the light path in the horizontal direction.

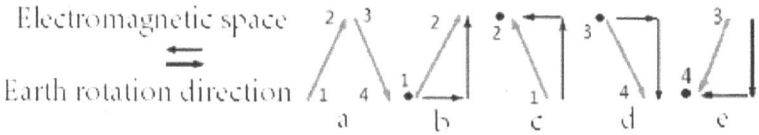

Fig. 1.3

At point 1 in the M2 optical path, this point represents the light emitting point of the light source. From the vector analysis, it can be seen that when the moving direction of the light source is consistent with the propagation direction of the light, the point will produce a blue shift.

Point 2 in the M2 optical path represents the receiving point of light. The direction vector of light is caused by the relative motion between the earth and electromagnetic space. The propagation direction vector of light is opposite to the moving direction of the earth, as shown in figure 1.3c. Since the point receives light, the movement of the point in the light facing direction also produces a blue shift.

Point 3 in the M2 optical path can be regarded as the light emitting point of the light source. As can be seen from Fig. 1.3d, the moving direction of the light source is the same as the propagation direction of the light, so the point will also produce a blue shift.

Point 4 in the M2 optical path represents the receiving point of light. The propagation direction of light is shown in figure 1.3e. Since the point is a light receiving point, the movement of the point in the direction facing the light will also produce a blue shift.

Through the above analysis, all four points in the M2 optical path will be blue shifted. After the light source passes through the spectroscope, not only the phase of the optical signal in the M2 optical path will be delayed, but also the wavelength of the light will change. Since M1 and M2 are perpendicular to each other, the experimental device cannot guarantee that the two beams have a fixed phase difference and the same frequency.

The interference of light should not only have a fixed phase difference, but also keep the wavelengths of two beams of light the same. In the double slit interference experiment of light, interference will occur as long as the light is a wave, because two beams of light are the same light source. In Michelson Morey's experiment, although the two beams of light also come from the same light source, they are perpendicular to each other. The rotation of the earth has the opposite effect on the two beams of light, and the wavelengths of the two beams will change in the opposite direction with the rotation of the experimental device. With the rotation of the experimental device, one light path produces a red shift, while the other light path produces a blue shift.

How to prove the above theory, just repeat the experiment. The Michelson Morey experimental device can rotate 360°. When the inclination of the two light paths and the inclination of the earth rotation direction are equal to a in Figure 1.4, M1 light path will have a red shift, and M2 light path will also have a red shift. When M1 and M2 optical paths have red shift or blue shift at the same time, and the frequency shifts of the two beams are equal, stable interference fringes can be observed on the observation screen.

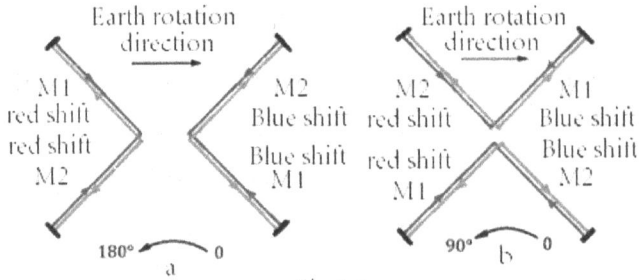

Fig. 1.4

If the experimental device is rotated again, one light path will have a red shift, while the other light path will have a blue shift. When the angle is rotated to b in Fig. 1.4, the reverse change of the two optical paths M2 and M1 reaches the maximum. The red shift of one beam reaches the maximum, while the blue shift of the other beam reaches the maximum.

Originally, the moving distance of the complete interference fringes can be seen when rotating 90° . However, due to the reverse change of the wavelengths of the two beams of light, when the experimental device is rotated again, the two beams of light cannot interfere with each other. What we saw in the experiment is that when the stable interference fringes are formed, if the equipment is rotated again, the interference fringes will not move, and the original fringes will also be disordered. However, when the experimental device is rotated 180° , the stable interference fringes can be observed again on the observation screen.

The results of Michelson Morey's experiment have a great impact on quantum mechanics. It not only changed the research direction of light, but also covered up the physical essence of space, and even hindered the development of the whole quantum theory. Due to the inherent defects of this experiment, the experimental results can neither deny the existence of ether, nor prove that the speed of light is constant.

§ 1·5 Two-way satellite time comparison between China and Japan

The principle of "constant speed of light" has its specific physical connotation. In "electromagnetic space", the speed of light propagation is a constant C. In geometric space, the speed of light is not necessarily constant C. "Electromagnetic space" is a physical concept, which has specific physical connotation. Geometric space is a mathematical concept, which reflects the geometric shape and spatial distance of space. Next, we will illustrate what is "electromagnetic space", what is geometric space, and what are the differences between them through a specific example. This can help us better understand the principle of "constant speed of light".

In order to improve the synchronization of world time, the International Bureau of Metrology (BIPM) proposed the global two-way satellite time transfer (TWSTT) plan. The Shaanxi Observatory of the Chinese Academy of Sciences (CSAO) and the Institute of communications of the Ministry of post of Japan (CRL) have carried out two-way satellite time signal transmission. The work has obtained relevant experimental data. The results show that the signal transmission time of the satellite over Tokyo from Japan to Xi'an in China is not equal to the signal transmission time of the satellite over Xi'an to the satellite over Tokyo.

Since Japanese satellites are over Tokyo and Chinese satellites are over Xi'an, they have the same latitude and are arranged according to the rotation direction of the earth. When the two satellites transmit time synchronization signals, the experimental properties are similar to the Michelson Morey experiment. The signal from Tokyo to Xi'an propagates in the opposite direction of the earth's rotation. The signal from Xi'an to Tokyo Travels in the same direction as the earth's rotation. If the propagation medium of light is "electromagnetic space", the rotation of the earth will cause the relative motion between the satellite and "electromagnetic space", which will produce the difference of two-way transmission time.

The Michelson Morey experiment did not get the expected results. However, the Sino Japanese satellite experiment clearly tells us that "electromagnetic space" exists. The relative motion between the satellite and the electromagnetic space will lead to the signal transmission between the satellite and the ground or between satellites, which may be transmitted along the direction of motion in the electromagnetic space or against the direction of motion in the electromagnetic space; this will not only cause frequency drift of the transmitted signal, but also affect the speed of signal transmission.

Some people believe that the error in the time calibration of China Japan two-way satellites is the "relativity" effect of relativity. Others deny relativity and doubt the principle of "constant speed of light". These two statements are caused by the incorrect understanding of the principle of "constant speed of light". The "relativity" effect of relativity is only an observational effect. The error in the time comparison between Chinese and Japanese satellites is the error caused by the relative motion between the satellite and "electromagnetic space".

Two satellites with different longitudes at the same latitude will have a relative motion with their surrounding "electromagnetic space" due to the rotation of the earth. According to the order of the earth's rotation direction, the East is the front and the west is the back. We can select the earth as the reference system. The movement direction of "electromagnetic space" relative to the earth is opposite to the earth's rotation direction.

If the eastern satellite sends a time comparison signal to the western satellite, when the signal sent by the eastern satellite enters the "electromagnetic space", the electromagnetic wave frequency will drift due to the earth's rotation, which is similar to the red shift of light. When the western satellite receives a signal from electromagnetic space, the frequency of the signal will also drift, which is similar to the blue shift of light. Since the relative velocity of the East and West satellites and the "electromagnetic space" is the same V, the "radial velocity" between the East and West satellites is zero. The frequency shifts generated during transmission and reception just offset each other.

Similarly, the West Satellite sends time comparison signal to the East Satellite. When the electromagnetic wave signal sent by the West Satellite enters electromagnetic space, the frequency will also drift, which is similar to the blue shift of light. When the East Satellite receives signals from electromagnetic space, the frequency of the received signals will also drift, which is similar to the red shift of light. Finally, the redshift and blue shift just offset each other. Although there is no frequency shift in the signals received by the East and West satellites, it is not that the signal frequency shift did not occur, but that the frequency shifts generated in the signal just offset each other.

Since the time comparison signal sent by the East satellite to the West satellite is the same as the moving direction of "electromagnetic space", the signal transmission speed from the East satellite to the West satellite is C+V. The time comparison signal sent by the West satellite to the East satellite is opposite to the moving direction of "electromagnetic space", and the signal transmission speed from the West satellite to the East satellite is C-V. The signal transmission speed takes "electromagnetic space" as the reference frame, and the signal transmission speed in "electromagnetic space" remains unchanged.

In the above process, the geometric space distance between the East satellite and the West satellite is equal to that between the West satellite and the East satellite, and the transmission speed of the signal in the "electromagnetic space" is also the same. However, the signal transmission time from the East satellite to the West satellite is different from that from the West satellite to the East satellite. This shows that the transmission space of time comparison signal is "electromagnetic space", not "geometric space". "Constant speed of light" is the speed of light propagation in "electromagnetic space", not the speed of light propagation in "geometric space". When there is no relative motion between geometric space and electromagnetic space, geometric space is equal to electromagnetic space. When there is relative motion between geometric space and electromagnetic space, geometric space is not equal to electromagnetic space. If the relative motion speed between "geometric space" and "electromagnetic space" is V, the speed of light propagation in "geometric space" is C+V or C-V. "Geometric space" is the space of distance between two communication satellites.

In the two-way satellite time comparison between China and Japan, due to the rotation of the earth, there is a relative movement between the "geometric space" from Xi'an to Tokyo and the "electromagnetic space" of electromagnetic waves, resulting in the transmission speed of the time signal from Tokyo to Xi'an is C+V, and the transmission speed of the time signal from Xi'an to Tokyo is C-V.

Some people believe that the two-way satellite time comparison between China and Japan measured the "relativity" effect of the earth's rotation, but did not measure the "relativity" effect of the earth's revolution speed. Or the "ether wind" of the earth's revolution has not been measured. Now let's analyze this problem.

To analyze the "relativity" effect of the earth's rotation, we can select the earth as the reference frame, and the earth's rotation can be regarded as the reverse movement of "electromagnetic space". To analyze the "relativity" effect of the earth's revolution, we need to select the Milky way as the reference system, because the earth's revolution includes the earth's revolution around the sun and the solar system's revolution around the Milky way. The selection of the Galactic reference system can reflect not only the motion of the earth around the sun, but also the motion of the solar system around the Milky way.

The background space of the Galactic reference system can be regarded as Newton's "absolute static space". Newton's "absolute static reference frame" is "electromagnetic space" in Min's space-time. Using Newton's "absolute static reference system" is to use "electromagnetic space" as the motion reference system. Under the background of "electromagnetic space" reference frame, the revolution of the earth is the superposition of two kinds of motion, and the motion trajectory is a spiral motion mode.

After using "electromagnetic space" as the motion reference frame, we don't need to consider the hierarchical structure of galaxies above the Milky way. The principles contained in this are the two conventions of Einstein's "special theory of relativity": the principle of relativity and the principle of "constant speed of light".

After considering the above problems, we can draw the "space-time map" of the earth's revolution. In Figure 1.5, A represents the position of the eastern satellite in "electromagnetic space", B represents the position of the western satellite in "electromagnetic space", the distance from 0 to B is the earth's revolution speed around the sun of 30km/s, and the distance from B to C is the earth's movement speed around the Milky way of 250km/s. When time t_1=1s, the western satellite will move from point B to point D. the connection between B and D is the trajectory of the western satellite under the Galactic reference system. Point A is the position of the East satellite in "electromagnetic space" when time t=0. Since the East and West are synchronous satellites, point D of the West satellite corresponds to point E of the East satellite.

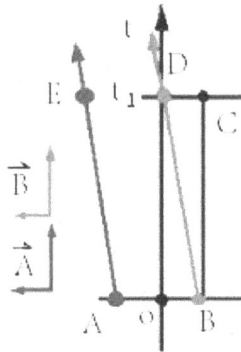

Fig. 1.5

In Figure 1.5, the direction vector of the West satellite motion is vector B, and the direction vector of the East satellite motion is vector A. In "electromagnetic space", both satellites have motion components of vertical "electromagnetic space" motion and horizontal motion.

As can be seen in Figure 1.5, the vertical motion component in the motion vector will not affect the two-way time information transmission of the East and West satellites, but only the horizontal motion component will affect the two-way time information transmission. When the eastern satellite sends time information to the western satellite, because the western satellite moves in the direction of signal transmission, it takes less time to transfer time information from the eastern satellite to the western satellite. If the western satellite sends time information to the eastern satellite, because the eastern satellite is moving in the opposite direction, the time for time information transmission from the western satellite to the eastern satellite will increase. Obviously, the earth's revolution will produce the "relative motion" effect.

The "relative motion" effect is caused by the relative motion between the satellite and the "electromagnetic space". The time error caused by the "relative motion" effect of the earth's revolution is a fixed value. The time error caused by the "relative motion" effect of the earth's rotation is also a fixed value. Since the earth's rotation and revolution directions are consistent, the relative motion direction relative to the "electromagnetic space" is also consistent. The relative motion has the superposition property, and the resulting error value also has the superposition property.

In the Sino Japanese satellite time comparison experiment, the "relative motion" effect of the earth's revolution was not observed, which is actually a mistake in understanding. The "ether theory" holds that the earth travels in the "Ether". If the rotation of the earth is moving along the "Ether" during the day, it is moving against the "Ether" at night. In this way, the earth's rotation will cause the "relativity" effect of the earth's revolution to change from $(+\Delta t)$—0—$(-\Delta t)$—0—$+\Delta t)$ The change process of. In the Sino Japanese satellite experiment, this change in the experimental data was not observed, which is considered to be that the "relative motion" effect of the earth's revolution was not observed.

The time comparison signal propagates in the "electromagnetic space", and the speed is a constant C. As long as there is relative motion between communication satellites and "electromagnetic space", the signal transmission time between communication satellites will be changed. Electromagnetic space is similar to Newton's absolute static space. Under the Galactic reference system, any movement of the earth, including rotation and revolution, is relative to "electromagnetic space". The moving reference object is the stationary space around the earth. For the geostationary space reference system around the earth, whether the earth rotates or revolves, there is no difference between anterograde and retrograde. (From this example, we can see the difference between "electromagnetic space" and ether) The earth's rotation and revolution cause error in signal transmission time, which is a fixed value. In the experiment of satellite time comparison between China and Japan, the measured error value is the superposition of two "relative motion" effects. The experimental data have included the errors caused by the earth's rotation and the earth's revolution. Since these two errors are fixed values, it is impossible to distinguish how much each is.

The axis of the rotation axis determined by the connecting line between the East and West satellites has an inclination with the earth's revolution plane. The angle of this inclination will change with the rotation of the earth. When the motion speed is converted into "radial speed", the converted "radial speed" will also change due to the change of the inclination angle of the earth's rotation axis. This change will cause the error value of the "relative motion" effect to fluctuate within a small range within 24 hours. The observation data obtained in the Sino Japanese satellite experiment have small fluctuations near a central value, which is caused by the change of the earth's rotation angle.

Take the Milky Way as the reference system, and stipulate that this reference system is Newton's "absolute static space"; therefore, the "electromagnetic space" is Newton's "absolute static space". This assumption has practical significance. The theoretical basis for doing so comes from two basic assumptions of Einstein's "special theory of relativity". Based on this assumption, we can use experimental methods to measure the speed of light propagation in space under the Milky Way reference system. If the speed is not 300000 Km/S, then the space under the Milky Way reference system is not "electromagnetic space", but "geometric space". If it is a "geometric space", we can correct it according to the principle described above and calculate the relative motion speed of the reference system. Only the speed of electromagnetic wave propagation in space is the speed of light, and this space is "electromagnetic space".

The experimental data of two-way satellite time comparison between China and Japan can be perfectly explained by using the Doppler shift of light, the principle of "constant speed of light" and the physical concepts of "electromagnetic space". This undoubtedly proves the correctness of the light theory mentioned above. Doppler redshift and the principle of "constant speed of light" are inseparable physical mechanisms. When there is relative motion in the reference system, either light will cause redshift, or light will change the propagation speed, which is called the "relative motion" effect.

The relative motion between "geometric space" and "electromagnetic space" will produce the "relative motion" effect. This "relative motion" effect will affect satellite communications, satellite networking, satellite ranging, satellite positioning, satellite navigation, etc. The relative motion speed of satellite and "electromagnetic space" is used to correct the speed of electromagnetic wave propagation. This method can calculate "time" and "space" more accurately. The physical principles can be applied in aerospace, aviation, autonomous navigation, etc. The application of relevant principles in "laser gyro" is discussed below.

§ 1·6 Relative motion effect and Sagnac effect

Whether the speed of light changes or does not change reflects the problem of whether there is relative motion between geometric space and "electromagnetic space", and it is also the relationship between Doppler redshift and "constant speed of light". This section discusses the relationship between these two mechanisms.

In Doppler redshift, "electromagnetic space" is a very important physical concept. It is different from "Ether". It is not a substance existing in space, but vacuum space itself. "Electromagnetic space" is not only the motion reference frame of electromagnetic wave, but also the carrier of electromagnetic wave. The relative motion between any electromagnetic radiation source and "electromagnetic space" will have physical effects. This physical effect is not limited to the range of visible light. For electromagnetic waves of any frequency, relative motion will lead to frequency drift of radiated electromagnetic waves.

When there is relative motion between the electromagnetic wave radiation source and the "electromagnetic space", if the motion direction is opposite to the propagation direction of the electromagnetic wave, this reverse motion will cause the electromagnetic wave length to become longer. When the electromagnetic wave radiation source moves in the same direction as the propagation direction of the electromagnetic wave, this same direction movement will cause the electromagnetic wave length to become shorter. The movement of electromagnetic radiation source causes signal frequency drift, which is a common phenomenon in non—geostationary orbit satellite communication.

Similarly, when the electromagnetic wave receiving device moves reversely against the electromagnetic wave propagation direction, or the electromagnetic wave receiving device moves in the same direction along the electromagnetic wave propagation direction, it will also shorten or lengthen the wavelength of the received electromagnetic wave signal. In the process of receiving and transmitting electromagnetic wave signals, frequency drift occurs. This physical phenomenon will occur in the process of satellite receiving ground signals or signal relay between satellites. Whether it is the electromagnetic wave signal transmitting source or the electromagnetic wave signal receiving device, the reference frame of their motion is "electromagnetic space".

If the relative motion of the electromagnetic wave signal transmitting device, the electromagnetic wave signal receiving device and the "electromagnetic space" is equal in speed, and the frequency drift caused by one is positive and the other is negative, the two can offset each other.

If the signal transmitting device and the signal receiving device have relative motion with the "electromagnetic space", but the received signal does not produce frequency shift. This shows that the geometric space distance between the signal transmitting device and the signal receiving device has not changed. There is no relative motion between the signal transmitting device and the signal receiving device.

When the geometric space distance between the light source and the observer does not change, but the "inertial system" and "electromagnetic space" where the light source and the observer are located have relative motion. At this time, the red shift or blue shift caused by relative motion will offset each other between the light source and the observer. The physical effect of relative motion between the "inertial system" and "electromagnetic space" where the light source and the observer are located will not disappear out of thin air, but will change the speed of light propagation in geometric space. At this time, the speed of light propagation in geometric space is no longer C, but C+V or C-V.

If the motion of the light source relative to the "electromagnetic space" produces a red shift. The movement of the observer relative to the "electromagnetic space" produces a blue shift. If the redshift and blueshift just cancel each other, there will be no frequency shift in the measured spectrum. But at this time, both the light source and the observer have relative motion with the "electromagnetic space", and this "relative motion effect" will be reflected in the speed of light propagation. At this time, the speed of light propagation in the geometric space between the light source and the observer is C + V or C-V.

In this way, we can define the "relative motion effect" produced by the relative motion between the light source or the observer and the "electromagnetic space": due to the relative motion between the light source, the observer and the electromagnetic space, this relative motion will produce Doppler red shift, and the Doppler red shift produced by the relative motion can keep the "speed of light unchanged". If there is relative motion between the observer and the electromagnetic space, but the light measured by the observer does not have spectral frequency shift, then in the "inertial system" where the observer is located, the speed of light is no longer C, but C+V or C-V. Where V is the relative motion speed of the "inertial system" and "electromagnetic space" where the light source and the observer are located.

The relative motion effect makes the transmission speed of electromagnetic wave signal behave as C+V or C-V. In fact, the speed of light here has not changed, but the optical path has changed, that is, the geometric space distance traveled by light has changed. The space for transmitting optical signals is "electromagnetic space". If the observer in face of the light propagation direction moves relative to the "electromagnetic space" and the geometric space distance of light propagation is L, then:

$$L=\Delta t*C-\Delta t*V \qquad 1—6$$

Here, L is the geometric space distance between the light source and the observer, C is the speed of light, and V is the relative motion speed between the observer and the "electromagnetic space", Δt is the light propagation time.

If the observer's along light propagation direction moves relative to the "electromagnetic space", then:

$$L=\Delta t*C+\Delta t*V \qquad 1—7$$

The specific application example of "relative motion effect" is the so-called "Sagnac effect". Sagnac effect is an annular "interferometer" invented by Sagnac in 1913. Let's talk about its principle. See Figure 1.6.

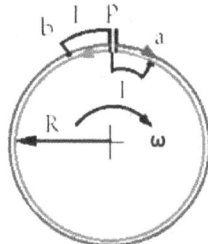

Fig. 1.6

A beam of light emitted from point P is decomposed into two beams, and they will meet after circulating in the same loop in the opposite direction. If the ring is stationary, the two beams of light reach point P at the same time. The "electromagnetic space" distance traveled by the two beams of light is equal to the geometric space distance. Geometric space distance $L=2\pi R$. R is the radius of the ring.

When the ring rotates clockwise at an angular velocity ω When rotating, the linear velocity of the ring is V, $V=R\omega$. The linear velocity V is the relative motion velocity between the geometric space determined by the ring and the "electromagnetic space". When the ring rotates clockwise at an angular velocity ω When rotating, the light going counterclockwise will coincide with the starting point at point a. The geometric space distance L_1 of light traveling counterclockwise is: $L_1 = L-I$. I is the arc length of the circle as it rotates. $I = \Delta t^*V$, Δt is the time taken for light to pass L_1.

When the counterclockwise light reaches point a, the clockwise light reaches point b. If the two beams of light converge to interfere, the geometric space distance of clockwise light is: $L_2=L_1+2I$, the optical path difference of the two beams is 2I. By detecting the optical path difference between the two beams, we can know the rotation angle and angular velocity of the ring.

Various "laser gyroscopes" made by using the principle of Sagnac effect have been widely used in aviation, aerospace, navigation and navigation and positioning of various vehicles. In particular, the "laser gyro" wound with optical fiber can change the sensitivity of the gyro through the number of turns wound with optical fiber. Optical path difference of one turn of optical fiber ΔL, ΔL is 2I, $\Delta L=2\Delta t^*V$.

The focus here is on the difference between geometric space and "electromagnetic space". The space of light propagation is "electromagnetic space", but light is refracted and reflected by optical fiber and bound in the geometric space constructed by optical fiber. The rotation of ring optical fiber is the movement of geometric space relative to "electromagnetic space". This relative motion will change the optical path of light in the geometric space of optical fiber.

The relative motion effect describes the physical effect of the relative motion between the light source, the observer and the "electromagnetic space". If the physical effect does not cause the Doppler redshift of light, it will change the speed of light in geometric space. If the light source and the observer are in the same reference frame and are in a relatively static state, the speed of light emitted by the light source will change in this reference frame. This principle can be applied in many fields, such as astronomical observation, space research, spacecraft, ballistic missile orbit design, measurement and control, autonomous navigation and so on. Here is a specific example to illustrate.

In addition to the "satellite positioning" system, some countries are studying the (MAPS) system. The (MAPS) system provides positioning, navigation and timing services when the "satellite positioning" system cannot work. (MAPS) is an autonomous navigation system. In addition to the general "strapdown inertial" navigation system, the system also has the ability to perceive and identify space-time information. How to make the autonomous navigation system obtain space-time information with coordinate characteristics in the process of operation? This is to turn the celestial reference frame (CRF) into a map of "space-time coordinates" with road signs.

"Earth Orientation" parameter (EOP) describes the earth's spatial orientation and rotation, which is an important connection parameter for the coordinate transformation between the celestial sphere and the earth reference frame. At present, there are three kinds of "celestial reference frame (CRF)", but they are not completely consistent in the specific implementation process. In fact, any reference frame will involve the selection of coordinate system. It is best to have an absolutely static reference system, which is the contradiction between Mach and Newton. Einstein has solved this problem. These are the two basic principles of "special relativity", the principle of relativity and the principle of "constant speed of light".

According to the two basic conventions of "special relativity", we can choose the Milky way as the moving reference system. The purpose of doing so is to take the background space of the Milky way as the absolute static space. When we take the background space of the Milky way as the motion reference frame, we can unify the celestial reference frame (CRF) and the earth's "orientation parameter" (EOP) in the coordinates with the Milky Way background space as the motion reference frame. When the "celestial sphere" and "Earth" use a unified coordinate system, it can bring convenience to aerospace, aviation, satellite measurement and control, and the implementation of autonomous navigation (MAPS) of aviation instruments.

Whether the background space of the Milky Way galaxy is an absolute static space needs to be confirmed by experimental data. The specific method is to measure whether the speed of light in space is constant C. If it is not constant C, it can be corrected by using the principle of "relative motion effect". The theoretical basis for measuring the speed of light is: absolute static space is "electromagnetic space", and the speed of light in "electromagnetic space" is C. If the speed of light is not C, the physical data of the relevant space needs to be corrected.

How to use experimental methods to verify whether the Galactic background space is an absolute stationary space? How to unify the celestial reference frame (CRF) and the earth's "orientation parameter" (EOP) in the Galactic background space? In fact, these are the same problem, and the principle of "relative motion effect" should be used. Use the celestial reference frame (CRF) for positioning and measurement, such as the sun orbit synchronous satellite, to establish a new celestial reference frame. The spatial direction and angular velocity of the earth's rotation are re measured by using the geosynchronous satellite. With the new earth "orientation parameters" (EOP) and celestial reference frame (CRF), we have a "space-time map" with road signs. In the (MAPS)autonomous navigation system, it can even design some operating parameters in the path in the orbit scheme.

When we determine the speed of light in the background space of the Milky way, we seem to get a ruler to measure the distance in space. This ruler can not only directly measure the spatial distance, but also analyze the spatial position and operating attitude of an aviation instrument in combination with the "relative motion effect", the principle of Doppler redshift, and the relationship between radial velocity and Doppler redshift.

Taking "electromagnetic space" as a unified reference standard not only solves the problem of coordinate conversion of different "inertial systems", but also enables the aviation instruments to read their own operation status information in real time. When the aircraft cannot receive the measurement and control signal from the ground, the aircraft itself can also conduct autonomous navigation according to the operation track, space direction, and operation attitude.

The direct image of "electromagnetic space" is the cosmic microwave background. In addition to the cosmic microwave background, "electromagnetic space" should also include the magnetic field distribution of the earth itself. The earth's magnetic field has not only directivity (compass), but also uneven distribution of magnetic field strength in space. The uneven distribution of the earth's magnetic field intensity is also an important space-time information. In addition to "electromagnetic space", there is also a "gravitational space" in space. The physical parameters of "electromagnetic space" and "gravitational space" can be used as important operating parameters of (MAPS) autonomous navigation system. (MAPS) autonomous navigation system should be equipped with more advanced "strap down inertial" navigation system and some special equipment. It not only detects some physical data of "electromagnetic space", but also measures some data related to "gravitational space".

The light we discussed above is related to "electromagnetic space", and the earth's gravity field is related to "gravitational space". Some physical parameters of the earth's gravity field are also important basis for time-space identification. Next, we will discuss "gravitational space". Let's first understand "gravitational space" from the cosmological red shift.

PART TWO

Cosmological redshift and space expansion

The red shift of light includes not only Doppler red shift, but also cosmological red shift. Cosmological redshift is caused by space expansion. In addition to discussing the difference between the two redshifts, this chapter also analyzes the relationship between cosmological redshifts and space expansion.

Doppler redshift is caused by the relative motion between the light source or observer and the "electromagnetic space". Therefore, the Doppler redshift reflects the relative motion speed between the light source, the observer and the "electromagnetic space". The cosmological red shift reflects the space distance between the light source and the observation. One is "motion speed" and the other is "space distance". The two redshifts reflect two different physical quantities.

Doppler redshift is the redshift generated during the energy exchange *between the light source and the "electromagnetic space", or the redshift generated when the observer obtains the light energy from the "electromagnetic space". Cosmological redshift is the redshift of light in the process of propagation in "electromagnetic space". The first kind of red shift is the conversion process of light radiation and reception, and the second kind of red shift is the propagation process of light.*

When light travels in "electromagnetic space", it has no physical connection with the light source or the observer. Whether the light source or the observer is in motion will not affect the physical characteristics of light propagation in space. When light travels in the "electromagnetic space", whether the light has a red shift is only related to the physical characteristics of the "electromagnetic space". Therefore, the cosmological red shift reflects the expansion of "electromagnetic space".

Doppler redshift and cosmological redshift completely describe the life course of "light". The two redshift theories complement and support each other. Doppler redshift and "constant speed of light" are two aspects of a physical mechanism. Cosmological redshift and Doppler redshift are different processes of light. Doppler redshift, "constant speed of light" and cosmological redshift constitute a complete theory of light. Any theory of light is correct only if it satisfies the three light theories at the same time.

§2·1 Cosmological redshift

Cosmological redshift is a redshift caused by space expansion, which is not only the conclusion of theoretical research and cosmic observation, but also a theory basically recognized by everyone. Why does space expansion cause the red shift of the spectrum? At present, there is not only a lack of system theory, but also the related concepts are vague.

Before understanding the cosmological red shift, we should first understand the concept of two spaces. As mentioned earlier, "electromagnetic space" is a vacuum space that carries and transmits electromagnetic waves. In addition to "electromagnetic space", there is another space in vacuum space, called "gravitational space". "Gravitational space" is the space area where mass matter interacts. "Gravitational space" consists of the gravitational action between celestial bodies. In this way, cosmic space or vacuum space can at least be divided into "gravitational space" and "electromagnetic space".

"Gravitational space" and "electromagnetic space" reflect the two forces contained in vacuum space, universal gravitation and electromagnetic force. The space area under the action of universal gravitation is "gravitational space", and the space area under the action of electromagnetic force is "electromagnetic space". There are two kinds of waves in space, electromagnetic wave and "gravitational wave". Electromagnetic wave is the vibration wave of electric field and magnetic field in "electromagnetic space". But "gravitational wave" is not a wave generated by gravity. (Note: "Gravitational wave" will be

specially discussed later)

The electromagnetic wave is transmitted in the "electromagnetic space", and its wavelength is related to the vacuum space carrying the electromagnetic wave. If the space expands, the transmitted electromagnetic wave will be stretched synchronously. Obviously, this view is based on the fact that there must be a carrier for the propagation of light, and this carrier is vacuum space; in this way, the wavelength of light will expand with the expansion of the space carrier.

Vacuum space is a physical entity, similar to an expandable balloon. After expansion, the space body not only stretches the electromagnetic wave length propagating in space, but also opens the space distance between the luminous source and the observer. The space between the light source and the observer is the space between two celestial bodies. The space between two celestial bodies is the "gravitational space" where gravity exists. Obviously, the expansion of space includes the expansion of "electromagnetic space" and the expansion of "gravitational space". The expansion of the two spaces is concentrated on the cosmological red shift of light, as shown in Fig. 2.1.

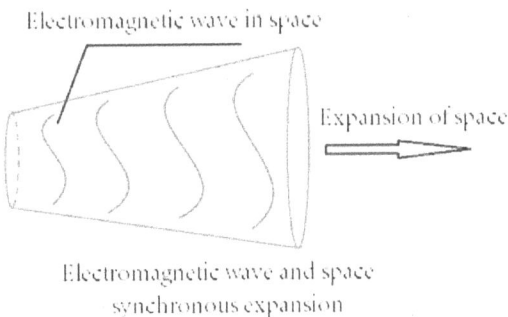

Electromagnetic wave in space

Expansion of space

Electromagnetic wave and space
synchronous expansion

Fig.2.1

Cosmological redshift reflects the expansion of "electromagnetic space" and "gravitational space" at the same time. The expansion of "electromagnetic space" determines the expansion rate of electromagnetic wave length. "Gravitational space" is the space distance traveled by light, which reflects the duration of electromagnetic wave length expansion.

if Z_C is the cosmological redshift, then:

$$Z_C = \frac{\Delta\lambda}{\lambda_0} \qquad 2—1$$

In it $\Delta\lambda$ is the amount of change in the wavelength of light, λ_0 is the original wavelength of light wave. If a beam of light is transmitted from celestial body a to celestial body b, the wavelength measured at celestial body b is λ, then:

$$\Delta\lambda = \lambda - \lambda_0 \qquad 2—2$$

If the expansion rate of "electromagnetic space" is τ, The time taken for light to pass from celestial body a to celestial body b is t, then:

$$\Delta\lambda = \tau^* t \qquad 2—3$$

The space between celestial bodies a and b is "gravitational space", and the size of t represents the spatial distance between the two celestial bodies. The expansion of space means that the space distance between the two celestial bodies is also increasing synchronously. The larger t is, the farther the space distance between the two celestial bodies is pulled. According to the above principles, cosmology redshifts Z_C : It not only reflects the expansion rate of "electromagnetic space", but also reflects the expansion rate of "gravitational space".

Cosmological redshift is the amount of wavelength change $\Delta\lambda$, And original wavelength λ_0 Ratio of. It can be seen from figure 2.1 that the change of electromagnetic wave length is synchronized with the expansion of space. Therefore, the cosmological redshift of the spectrum Z_C is also equal to the relative expansion rate of space.

The equivalent relationship between electromagnetic wave length and spatial expansion not only proves that the cosmological redshift is caused by spatial expansion, but also determines the corresponding relationship between the value of redshift and the rate of spatial expansion. Moreover, this correspondence is not limited to the electromagnetic wave in the visible light range, but the whole electromagnetic wave spectrum.

If the scale factor a(t) is used to represent the size of cosmic space, $a_{(now)}$ is the current value (when receiving the optical signal), $a_{(then)}$ is the value of the past (when the luminous object emits light), then:

$$\frac{a_{(now)} - a_{(then)}}{a_{(then)}} = \frac{\lambda - \lambda_0}{\lambda_0} \qquad 2\text{—}4$$

According to this equivalence relationship, the parameters of space expansion can be used to calculate the cosmological redshift, and the data of redshift can also be used to calculate the rate of space expansion.

Cosmic space scale factor is a parameter of Friedman equation and a time function of the relative expansion of the universe. The change of scale factor reflects the change of cosmic space. According to the equivalence relationship of formula 2-4, the value of cosmological red shift can be directly obtained from the ratio of the increase of scale factor to the original value:

$$Z_C = \frac{a_{(now)} - a_{(then)}}{a_{(then)}} \qquad 2\text{—}5$$

Normalize the scale factor $a_{(now)}$ of the present tense to 1, then:

$$a_{(now)}=1 \qquad \frac{1}{a_{(then)}}=Z_C+1 \qquad\qquad 2\text{—}6$$

The above formula can connect the redshift of spectrum with the relative expansion speed of space, and use the cosmological redshift to calculate the expansion speed of cosmic space.

In the cosmological red shift, $\Delta\lambda$ It involves two spaces, namely electromagnetic space and gravitational space. The expansion rate refers to the expansion rate of the "electromagnetic space" carrying electromagnetic waves. The "electromagnetic space" is a flat and straight "linear space" that fully conforms to the characteristics of Minkowski Space time. The second derivative of the scale factor $a(t)$ in this space is equal to zero. Therefore, the expansion rate of electromagnetic space is a constant.

Time t describes the space distance traveled by light, which is the space between the observer and the luminous celestial body, and the "gravitational space" between the two celestial bodies. "Gravitational space" is a nonlinear space, and the second derivative of the scale factor of the space is not zero. Therefore, the cosmological redshift formula is a simple approximate algorithm obtained by replacing the scale factor of "gravitational space" with the scale factor of "electromagnetic space". Under the large-scale structure of the universe, when the redshift is less than 1, it has a high degree of approximation.

Approximate algorithm, omitted $\Delta\lambda$ the integral solution of value avoids the integration of time or path. If you want to calculate accurately $\Delta\lambda$ the field equation of "general relativity" needs to be used to calculate the light line distance.

§ 2·2　The physical nature of space

When we discuss Doppler redshift, we introduce the concept of "electromagnetic space", and when we discuss cosmological redshift, we have the concept of "gravitational space". Now we need to introduce a new concept of space, which is "flat and straight space". These concepts of different spaces are very important physical words, which are the basic knowledge we must master to understand the universe. The introduction of these concepts of space is to better understand the largest existence in the universe, which is called cosmic space.

We have many reasons to believe that vacuum space is a physical entity. "Electromagnetic space" and "gravitational space" are physical terms reflecting certain physical properties of vacuum space. Similarly, "flat and straight space" is also a physical vocabulary reflecting some functional characteristics of vacuum space.

The original physical concept of vacuum space is: three-dimensional open cavity without any substance. This concept regards space as a void without any substance. There is no matter in the vacuum space should be understood as: vacuum space refers to the pure space body excluding any ordinary matter. Then, except for ordinary matter, all cosmic space is vacuum space. Vacuum space is not a void, but another form of matter. Vacuum space is the basic space of other physical spaces. Other physical spaces reflect the physical characteristics of vacuum space from different aspects.

The first thing that reflects the physical characteristics of vacuum space is "flat and straight space". "Flat and straight space" reflects the original state of vacuum space. If there is no mass matter in the vacuum space, then the vacuum space is a flat and straight space. The feature of "flat and straight space" is flat and straight, which conforms to the feature of Minkowski Space-time 3D space. Minkowski Space-time is the simplest pseudo Riemannian space, satisfying that Riemannian curvature is equal to zero.

The flatness and straightness of a vacuum space do not lie in its shape, but in its physical function. "Flat and straight space" expresses the flat and straight characteristics of vacuum space. Minkowski Space time is the background space of "Special Relativity", because the theory of "Special Relativity" is based on "flat and straight space". The reason why "electromagnetic space" is the reference space of light speed is that "electromagnetic space" inherits the flat and straight characteristics of vacuum space. Light propagates linearly in vacuum space because "electromagnetic space" also has the flat and straight characteristics of "flat and straight space".

The flatness and straightness of space is a physical property, which represents a function of space. If Newton's formula of universal gravitation is not supported by the flatness and straightness of space, the conclusion that the force is inversely proportional to the square of distance cannot be held. "Flat and straight space" not only defines the straightness of dimension and the uniformity of scale, but also determines the uniformity of time process. Vacuum space has these characteristics of "flat and straight space", which also lays a space foundation for "electromagnetic space" and "gravitational space".

Electromagnetic space refers to the space area where vacuum space transmits and carries electromagnetic waves. "Electromagnetic space" reflects the electromagnetic characteristics of vacuum space. "Electromagnetic space" has its specific spatial connotation, and the "microwave background" picture of the universe is the real picture of "electromagnetic space". "Electromagnetic space" does not only refer to the range of visible light, but includes all electromagnetic wave spectrum. If light fills all the universe, then the scope of vacuum space is the scope of "electromagnetic space". "Electromagnetic space" expands with the expansion of vacuum space. The wavelength of electromagnetic wave also expands with space in a specific proportion. The cosmic microwave background image records the changing process of electromagnetic space.

According to the big bang theory, the age of the universe when photons are "decoupled" is about 3.8×10^5 years, and its diameter is about 2.55 million light years. Now the universe is about 92 billion light-years in diameter, expanding by more than 3×10^4 times. If the wavelength of electromagnetic wave and the range of space are increased in the ratio of 1:1, the visible light with the original wavelength of 5×10^{-7}m has become the microwave with the wavelength of about 1.8×10^{-2}m. That is to say, if this proportional relationship holds, the diameter of the universe is more than 3×10^4 times larger than when the universe was only 3.8×10^5 years old; the diameter of "electromagnetic space" has also increased by more than 3×10^4 times; the electromagnetic wave length during the big bang period also increased by more than 3×10^4 times.

In addition to the straightness of vacuum space and the same spatial expansion rate, "electromagnetic space" also has isotropy and uniform spatial distribution. "Electromagnetic space" has the same spatial characteristics as vacuum space, which is very useful for cosmic research. Some physical parameters of vacuum space cannot be measured, while the spatial characteristics of "electromagnetic space" can be observed and measured. In this way, we can use some physical properties of "electromagnetic space" to study vacuum space. For example, the bending of light near a massive object is actually caused by the bending of the vacuum space around the massive object. For another example, "electromagnetic space" and vacuum space have the same spatial characteristics, which shows that "electromagnetic space" can be used as Newton's absolute static space. In astronomical observation and "MAPS" autonomous navigation system, the "microwave background" radiation, including some parameter characteristics of electromagnetic radiation in various frequency bands of the Milky Way, can be directly used as the operating parameter index or motion reference of space vehicles.

Gravitational space is a space area where mass matter interacts. What is the relationship between mass matter interaction and space? Why use space to describe gravity? We all know that gravitation is a force described by Newton and possessed by all things, but Einstein did not think so. Einstein believed that gravity is not the function of matter, but caused by space bending.

Vacuum space is originally a flat and straight space. When there is mass matter in the space, the mass matter will destroy the original flatness and straightness of the space. Flat and straight is a characteristic of vacuum space, similar to the elastic effect of elastic materials. When a "foreign body" is inserted into the elastic material, the elastic material will deform, and the stress generated by deformation will repel and compress the "foreign body". Mass matter in vacuum space will also cause space deformation. After space deformation, it will also produce a repulsion, which is manifested as compression of this mass matter. The compression of space on this mass material is as if there is gravity in the material itself. If the mass material is composed of two particles, space will regard the two particles as a whole. Space compresses the whole composed of two particles, as if there is gravity between the two particles.

Gravity is a compression force in space, which is caused by the deformation of space. The deformation of space is caused by the mass of matter. Newton described the effect of mass of matter and called it universal gravitation. Einstein's field equation describes the deformation of space, which generates gravity. Newton and Einstein complement each other, which can perfectly explain the essence of gravity.

"Gravitational space" is the whole universe we see, which is composed of stars and celestial bodies. We originally thought that they were connected by gravity, but actually they were formed as a whole by the repulsion and compression force of space. "Gravitational space" is a space area where vacuum space is deformed. When space expands, "gravitational space" expands with it. The expansion of "gravitational space" cannot be simply understood as the separation of galaxies. According to the viewpoint of "general relativity", the expansion of "gravitational space" should be understood as that curved space-time tends to be flat and straight.

Flat and straight space reflect the morphological characteristics of vacuum space, electromagnetic space reflects the electromagnetic characteristics of vacuum space, and gravitational space reflects the deformation characteristics of vacuum space. Vacuum space is the basic space of all spaces. Energy forms "electromagnetic space" through electromagnetic action in space. Mass changes the flatness and straightness of space, forming a "gravitational space". "Electromagnetic space" is the space range of energy action, and "gravitational space" is the space range of mass action. The effects of electromagnetic force and universal gravitation form two different spaces.

According to cosmological redshift, space expansion is the simultaneous expansion of "electromagnetic space" and "gravitational space". Because vacuum space is the basic space of "electromagnetic space" and "gravitational space". Therefore, we can easily come to a conclusion that cosmological redshift reflects expansion of the vacuum space.

The expansion of vacuum space is the expansion of "flat and straight space", which is consistent with the observation that the universe is flat. In the theory of space expansion, we should also discuss why space expands; the expansion of space is an acceleration expansion. Obviously, these problems involve "dark energy". In the analysis of Doppler redshift, the relevant mechanism of the principle of "constant speed of light" is solved. In the analysis of cosmological redshift, we come to the conclusion that space is expanding. When analyzing the physical nature of space, we hope to solve the mystery of "dark energy". Before doing so, we need to understand Einstein's "principle of equivalence".

§2·3 Weak "equivalence principle"

Einstein had considered the difference between "gravitational space" and "flat and straight space" when he put forward "general relativity". In his thought experiment of "equivalence principle", there are two different experimental environments for spaceships: one is on the surface of a planet, and the spaceship is in the field strength of a gravitational field; the other is that there is no star or gravity field in the deep space or nearby; these two different experimental environments refer to "gravitational space" and "flat and straight space".

The "equivalence principle" in "general relativity" is that "gravitational mass" is equivalent to "inertial mass". In fact, there are several problems that physics has not yet figured out. One is why all substances have "mass", and how does "mass" come into being? The other is what is "inertia"? What is the relationship between "gravitational mass" and "inertial mass"? When we have the concept of different spaces, these problems become simple.

"Mass" is an interaction between matter and surrounding space. There are two kinds of "mass", "gravitational mass" and "inertial mass". "Gravitational mass" refers to the interaction between matter and the surrounding "gravitational space"; "inertial mass" refers to the interaction between matter and surrounding "vacuum space".

According to the law of universal gravitation, "gravitational mass" is a measure of the interaction between this particle and another particle. Considering the influence of gravitational constant G and other particles in space. If a particle is in space and there are other planets or gravity fields around it, and other planets or gravity fields constitute the environment of the whole space of the particle, the action object of the particle is the whole "gravity space" around it.

"Gravitational mass" is a measure of the interaction of particles in "gravitational space". If there is no planet or gravity field around the space where the particle is located, the interaction between the particle and the surrounding space is "inertial mass".

Therefore:

"Gravitational mass" is a measure of the interaction between particles and the surrounding gravitational space in the "gravitational space", expressed in m_g.

"Inertia mass" is a measure of the interaction between particles and the surrounding vacuum space in the "vacuum space", expressed in m_a.

Obviously, "inertial mass" is a measure of the interaction between a particle and its surrounding "vacuum space" (its principle is described later). Some people may think that there is no mechanical effect between particles and their surrounding "vacuum space" in the "vacuum space". How to identify that they have mechanical force. The effect of "vacuum space" on moving particles is a normative force, which is manifested as: static is constant static, and dynamic is constant motion. When a particle in vacuum space wants to change its motion state, the particle will be subjected to a reverse force that prevents its motion. The magnitude of the reverse force is related to the acceleration of the particle velocity. The larger a is, the greater the reverse force the particle receives.

In Figure 2.2, if the spacecraft accelerates with acceleration a in "vacuum space", particle m will receive a reverse force F_a, F_a is the force acting on particle m by "vacuum space" to prevent particle m from accelerating. If g is the gravitational acceleration of a particle in a gravitational field, it can be proved that when $g = a$, $F_g = F_a$.

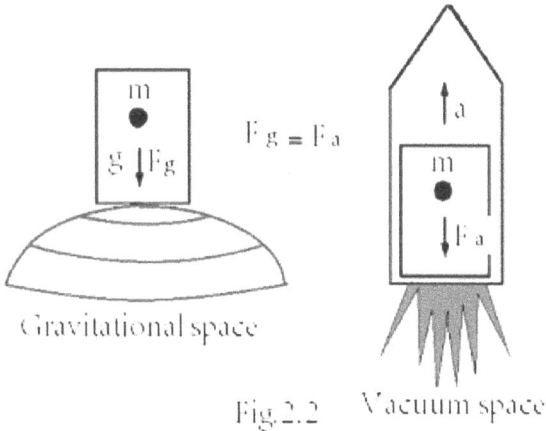

Fig. 2.2

A particle moving with acceleration a in "vacuum space" is equivalent to being in a gravitational field with gravitational acceleration g . On the contrary, in the gravitational field, the spacecraft accelerates g by gravity to move along the direction of the gravitational field. At this time, the space environment felt by particle m is equivalent to that felt by particle in vacuum space.

This "equivalence relationship" reflects the relationship between "gravitational space" and "vacuum space", and the essence of the two spaces is the same. In "gravitational space", the acceleration motion of particles can counteract the gravitational effect of gravitational field. The acceleration motion of particles in "vacuum space" can create the gravitational environment in "gravitational space".

The above can be summarized as follows:

1. The essence of gravitational space and vacuum space is the same. The difference between gravitational space and vacuum space is the existence and absence of matter.

2. In gravitational space, a particle moving with acceleration along the gravitational field, the space environment it feels is equivalent to the space environment it feels in vacuum space.

3. In vacuum space, when a particle moves with acceleration, the particle will be subjected to the same interaction force as the gravity field. The vector direction of this interaction force is opposite to the direction of motion, and the magnitude of the force is directly proportional to the acceleration a.

4. A particle at rest in the gravitational field is equivalent to a particle in accelerated motion in vacuum space.

When a particle is stationary in the gravitational field, the force on the particle is F_g. When a particle accelerates in a vacuum, if the acceleration is a, the force on the particle is F_a. The "gravitational mass" of the "equivalence principle" is equivalent to the "inertial mass". In addition to $m_g = m_a$, it also shows that when $a = g$, $F_a = F_g$. F_a and F_g are two forces with different properties. F_g is the force to accelerate the movement, and F_a is the force to prevent the accelerated movement. The "equivalent principle" reflects the force of particles in the two spaces, and also reveals some physical characteristics of the "vacuum space". The vacuum space here is flat, straight space, or the depths of the universe without any stars or gravitational fields nearby.

PART THREE

Flat and straight space and gravitational space

The "equivalence principle" discusses the relationship between "gravitational space" and "vacuum space" from the relationship between "gravitational mass" and "inertial mass". Vacuum space is the space without stellar matter, and "gravitational space" is the space with stellar matter. The difference between "vacuum space" and "gravitational space" is whether it contains matter.

"Gravitational space" refers to the vacuum space containing stellar matter, so the whole universe we observe is "gravitational space". We observe that the universe is expanding, which actually means that the "gravitational space" is expanding. But the cosmological red shift tells us that the expansion of "gravitational space" is due to the expansion of vacuum space; the expansion of vacuum space leads to the expansion of "gravitational space". The vacuum space is expanding, if we invert time, *when time returns to zero, the vacuum space is an infinitesimal point.*

The original "vacuum space" is an infinitesimal point. What does this mean? This relates to the origin of the universe; now the boundless space was created after the Big Bang. The Big Bang did not produce so much stellar material from nothingness, but blew a transparent hole out of an original chaotic state, and the hole is still expanding. This hole is the vacuum space. Obviously, this is consistent with the expansion of the present vacuum space.

The Big Bang theory is now a relatively accepted theory. The theory holds that the present universe is formed by a big explosion of a dense, hot singularity. However, this theory is flawed. In order to make up for this defect, some irrelevant hypothesis models are added to the theory. If the Big Bang is regarded as the explosion of space, all problems can be solved. Obviously, the understanding of the Big Bang as an expansion of space is inspired by the cosmological red shift.

§ 3·1 Flat and straight space and dark energy

The main body of the Big Bang is vacuum space. Vacuum space is the basic space of all spaces. Its original physical characteristics are flat and straight, so it is also called flat and straight space. The flat and straight space is a vacuum space without any mass matter. We can remove the mass matter and use the Friedman equation to describe this "flat and straight space".

Friedman equation is a set of equations that describe the universe model of uniform, isotropic and expanding space under the framework of "general relativity". As long as we remove the mass of matter from the equation and the space curvature is zero, the universe described by the equation is flat and straight space. In Friedman's equation, the material term $\rho = 0$, curvature $\kappa = 0$ Therefore, the equation:

$$H^2 = (\frac{\dot{a}}{a})^2 = \frac{8\pi G}{3}\rho - \frac{kc2}{a} + \frac{\Lambda c^2}{3}$$

Change to:

$$H = \frac{\dot{a}}{a} = \sqrt{\frac{\Lambda c^2}{3}}$$

3—1

Formula 3-1 shows that in "flat and straight space", the Hubble parameter is a constant $H = H_0$ The first derivative of scale factor a(t) is also constant. The first derivative of a(t) is a constant, which means that the space is a process of uniform expansion. The size of a(t) represents the size of cosmic space, which is only related to time. Normalize the current scale factor a(t) to $a_{(now)} = 1$. When t = 0, a(t) =0 and space is zero. In Figure 3.1, coordinate Y represents the value of scale factor a(t) , and X represents cosmic time. ρ Represents the density of matter, and ρ_{CRIT} represents the critical density of matter. $\rho/\rho_{CRIT} = \Omega(m)$ represents the universe with mass of matter. $0/\rho_{CRIT} = \Omega(0)$ represents flat and straight space without ordinary substances. In this way, we can get the a(t) diagram of flat and straight space, as shown in Figure 3.1:

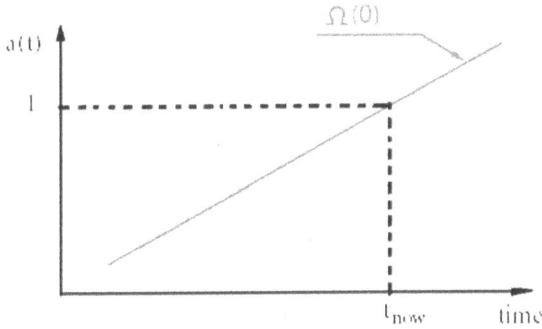

a(t) graph of flat and straight space

Fig. 3.1

Figure 3.1 shows that "flat and straight space" is a space expanding at a uniform speed. $\Omega(0)$ represents the overall behavior of the universe without ordinary matter. The "flat and straight space" in the expansion movement is similar to a train running forward. Stellar matter is only the passengers in this train. The behavior of the train constitutes the reference standard for all material movements in the universe. Stellar matter that expands synchronously with "flat and straight space" is considered to be in an absolute static state. On the contrary, stellar matter that is not in synchronous expansion motion with "flat and straight space" is considered to be in absolute motion.

The "equivalence principle" tells us a basic law of the operation of the universe: the motion with "flat and straight space" as the reference standard, whether accelerating or decelerating, is equivalent to the gravitational action of a reverse gravitational field, that is:

$$m_a a = -F_a \qquad\qquad 3\text{—}2$$

In the formula, m_a is the "inertial mass" of the particle and a is the acceleration of the relative motion between the particle and the flat and straight space. F_a is the force on the particle in the "flat and straight space". A minus sign indicates that this is a force that prevents movement. Formula 3-2 is exactly one more minus sign than Newton's second law.

The above principle shows that the synchronous expansion of stellar matter and "flat and straight space" is equivalent to being in an "absolute static" state; the relative motion between stellar matter and "flat and straight space" is "absolute motion". Absolute motion is limited by the speed of light. The expansion of stellar material in synchronization with the "flat and straight space" is equivalent to being in an "absolute static" state. Therefore, the expansion of space is not limited by the speed of light. This principle explains why space can expand faster than the speed of light. Therefore, the phenomenon that the expansion speed of the universe exceeds the speed of light can also prove that the superluminal movement between galaxies separated from each other can only be caused by the expansion of the space itself, rather than driven by other dark energy.

The physical characteristics of "flat and straight space" indicate its subject identity. In the first chapter, why take "vacuum space" or "electromagnetic space" as the reference frame of the speed of light. This paper not only explains the reason, but also points out the physical mechanism of the uniform expansion of "flat and straight space". The theory of space expansion of "flat and straight space" can solve the original debate about motion reference frame by Newton and others. At the same time, it also improves the theoretical framework of the principle of "constant speed of light". "Flat and straight space" is not only the motion reference frame of the speed of light, but also the reference standard of all physical behaviors in the universe.

The motion of stars and celestial bodies: one is the relative motion within and between galaxies, and the other is the separation between galaxies with the expansion of space. When we choose any inertial reference frame, the motion we observe is the superposition of the two motion quantities. The observed redshift is also the superposition of two different redshifts. The principle is very important in the study of the universe.

The galaxies around us are moving away from us, and the farther away they are, the faster they retreat, which is the result of the overall expansion of space. The expansion of "flat and straight space" is the overall behavior of the universe, accounting for about 70% of the motion energy of the universe, while mass matter accounts for only 30%.

In cosmology, the energy that dominates the expansion of space is called "dark energy". There are two reasons: first, the "flat and straight space" is not regarded as a physical existence, which makes this energy lose its physical support. Second, the observation of type Ia supernovae shows that the universe is not only expanding, but also accelerating. This additional energy seems to be unable to be explained by the big bang theory. Next, we will analyze the operation law of gravitational space and the physical mechanism of the accelerated expansion of the universe through the a(t) diagram of gravitational space.

§ 3·2　Gravitational space and the physical mechanism of the accelerated expansion of the universe

Here we first regard gravity as a force possessed by matter itself. Gravitational space is the space range of interaction of material forces. In gravitational space, in addition to participating in the interaction of universal gravitation, celestial bodies are also subject to the regulation of flat and straight space. In "gravitational space", in addition to participating in the expansion of "flat and straight space", the movement of celestial bodies also has its own laws of motion. Within the framework of material structure, because its four basic forces are greater than the expansion force of "flat and straight space", the material structure of atoms, molecules and aggregates will remain stable. In the structure of galaxies and star clusters, if the gravitational force between celestial bodies is greater than the expansion force of "flat and straight space", the structure will remain unchanged. Otherwise, the space between them will expand with the expansion of "flat and straight space". The structure, size and shape of "gravitational space" are also related to the formation, distribution and movement of stellar objects.

After the big bang, among the four basic forces formed in space, the weak force is the first, followed by the strong force and electromagnetic force, and finally the universal gravitation. The cosmic "microwave background" radiation shows that 380000 years before the formation of the universe, matter filled the whole cosmic space evenly in ionic state, and there is no gravitational space at this time. Strong interaction produces baryons, and electromagnetic interaction forms neutral atoms and molecules, and then to the aggregation state of matter. It will be a long time before "gravitational space" is built by stars and celestial bodies.

The "gravitational space" constructed by stellar matter is an important part of cosmic space. After the big bang, on the one hand, "gravitational space" should expand synchronously with "flat and straight space". On the other hand, due to the action of gravity, "gravitational space" has its own law of motion. Figure 3.2 is the a(t) diagram of "gravitational space" formed on the basis of "flat and straight space". The red line $\Omega(m)$ represents the gravitational space constructed by stellar matter.

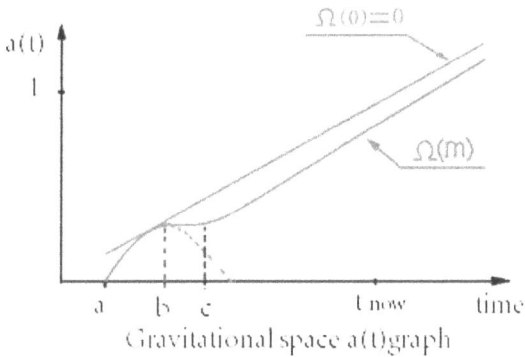

Gravitational space a(t)graph

Fig. 3.2

Between points a and b on the X time axis is the stage from the formation of the mass of matter to the expansion of the "gravitational space". In the cosmic microwave "background radiation" diagram, it is the process of forming the large-scale structure of matter due to the disturbance of small temperature difference. The Big Bang is the expansion of space. When space was formed, matter filled space with the embryonic form of fermions. At this time, the matter in space does not generate gravity. When the space expands and the temperature drops to a certain extent, the material can form a structure, and the material forming the structure first produces mass. The mass effect forms gravitational space in space. Between points a and b on the X axis, it not only reflects the process of matter from producing mass to forming gravitational space, but also reflects the initial stage of gravitational space starting to follow the flat and straight space and expanding simultaneously. Gravitational space expands at the same time as flat and straight space, which is also a process in which kinetic energy is converted into potential energy after matter forms the celestial structure.

Point b is the turning point from kinetic energy to potential energy. At this time, the material density in space, $\Omega(m)=\rho/\rho_{CRIT} \geqslant 1$. Due to the action of universal gravitation, "gravitational space" begins to shrink under the action of gravitation. At this time, the expansion speed of "gravitational space" is different from that of "flat and straight space".

When a particle M in the expansion space expands synchronously with the space, the particle M is in an unstressed state. If the particle M moves against the expansion motion of space with acceleration g under the action of universal gravitation, it means that the particle M is not synchronized with the expansion motion of space. At this time, particle M is not only affected by universal gravitation, but also by the force of space expansion motion. Gravitation is expressed in F_g. The force of space expansion is expressed by F_a. F_g is given by Newton's formula of universal gravitation. F_a is given by formula 3-2.

Formula 3-2 $F_a=m_a a$. m_a Is the "inertial mass" of particle M and a is the acceleration. Acceleration a is the velocity of particle M relative to the surrounding space, which is "flat and straight space". At this time, whether the particle M tends to expand or contract depends on the size of F_g and F_a.

Three situations: $F_a < F_g$. $F_a = F_g$. $F_a > F_g$.

If $F_a < F_g$: the gravitational space will shrink completely, and the cosmic operation mode will enter the red dotted line in Figure 3.2. Such a universe is an oscillating universe, and "gravitational space" circulates between explosion and contraction. When the universe enters this cycle mode, it is difficult for matter to evolve into advanced life forms and develop intelligent creatures such as human beings in a cycle of life. The development of our universe today shows that "gravitational space" does not enter the motion mode of red dotted line.

When $F_a = F_g$: the gravitational action of particle M in "gravitational space" is F_g, and it will contract at the motion speed of acceleration relative to "gravitational space". Since the expansion rate of "flat and straight space" remains unchanged, when the particle M shrinks with the acceleration of g, it is equal to the opposite motion with the acceleration of a relative to "flat and straight space". According to formula 3-2, particle M will be subjected to the reverse force exerted by flat and straight space: $-F_a = m_a a$. Since g = a, $-F_a = m_a g$, $F_a = F_g$, the total resultant force on particle M is zero. At this time, gravitational space can neither expand nor contract.

If $F_a = F_g$:This is an unstable structure, which will be broken by gravitational disturbance or space fluctuation. If the gravitational disturbance causes the "gravitational space" to shrink. "Gravitational space" shrinks → space radius decreases → gravity increases → space shrinks further. Obviously, this is a positive feedback process, which will make the "gravitational space" enter the motion state indicated by the red dotted line in Figure 3.2. This is the motion mode of the universe $F_g > F_a$.

If gravitational disturbance or spatial fluctuation makes $F_a > F_g$, the "gravitational space" will develop in the direction shown by the red line. This process is also a positive feedback process. As long as F_a is slightly greater than F_g, "gravitational space" will expand with the expansion of "flat and straight space". Space expansion will increase the radius and reduce the gravitational effect. When this positive feedback process is further deepened, "gravitational space" will accelerate and expand for the second time.

The red b-c line in Figure 3.2 indicates that the "gravitational space" is in the process of positive feedback. Point c on the red line is not only $\Omega(m) < 1$, but also a particle M, the acceleration a obtained is greater than the "escape velocity", i·e. $a > (2GM/D)^{1/2}$. Therefore, after the "gravitational space" passes point C, there is a stage of accelerated expansion.

Point c is the turning point of the expansion rate of "gravitational space". After point c, the "gravitational space" tends to be synchronized with the expansion of "flat and straight space". If we look behind point c, galaxies in this period will accelerate their expansion. This is why type Ia supernovae are observed to accelerate their expansion.

The essence of this accelerated motion of galaxies is to synchronize with the expansion of the surrounding physical space. If the physical space where the galaxy is located is selected as the reference object of motion, the accelerated expansion of the galaxy is a kind of motion tending to be static. This situation is similar to the free landing of an object in a gravitational field.

Through the above analysis and the principle shown in the a(t) diagram of "gravitational space", the "gravitational space" can only infinitely approach the expansion speed of "flat and straight space", but cannot exceed this speed. "Flat and straight space" relative to "gravitational space" is similar to Newton's absolute static space.

Einstein's field equation describes "gravitational space", and the field equation is:

$$R_{\mu\nu} - \frac{1}{2} R g_{\mu\nu} = k T_{\mu\nu}$$

The left side of the equation represents the curvature of gravitational space, and the right side of the equation represents the distribution of physical quantities such as mass, energy and pressure in space. The physical meaning expressed by the equation is that the basic space established by the equation is a flat and straight space. The existence of material mass in the space will lead to space bending. There is a corresponding relationship between space bending curvature and physical quantities such as mass, energy and pressure. The equation expresses this relationship.

If space is regarded as a physical entity, the field equation becomes very easy to understand. Due to the existence of material mass, the flexible, originally flat and straight space body will be deformed. The deformation of an object will produce stress inside, and the object inside the deformation will be subjected to deformation stress, which is the source of gravity.

It can also be seen from Figure 3.2 that if there is no expansion repulsion generated by flat and straight space in the universe, the "gravitational space" will enter the trajectory shown by the red dotted line, so that the universe will oscillate back and forth between contraction and expansion. In order not to let the universe oscillate back and forth, Einstein added a constant term to the field equation. Einstein's field equation with constant term:

$$R_{\mu\nu} - \frac{1}{2} R g_{\mu\nu} + \wedge g_{\mu\nu} = k T_{\mu\nu}$$

The $\wedge g_{\mu\nu}$ in the equation is the cosmological constant term. The cosmological constant term actually represents the repulsive metric tensor of the expansion of "flat and straight space". Einstein believed that adding a constant term to his field equation was the biggest mistake of his life. It is this error that makes up for the defect of the field equation.

PART FOUR

The main body of the world

Vacuum space is also a physical entity. The shock brought by this conclusion is huge. It subverts the original concept of matter. Matter can not only constitute the sun, moon and stars that we can see, but also exist in a physical way that we cannot perceive. We need more facts and reasons to prove this conclusion.

The theory described above involves two aspects: one is "light", the other is "space". "Light" is the medium for human beings to understand the world, and space is the basis for the existence of the world. If the theory described above is correct, it will impact the original physical theoretical system. As far as the current theory of "light" and "space" is concerned, the understanding of both is superficial and does not touch the essence of things. We need to thoroughly understand the essence of "light" and "space" in principle.

Vacuum space and the sun, moon and stars are not the same material form, and matter can exist in different forms. So, what is matter? When we still don't know much about this problem, we need to use some philosophical thinking methods. Some theoretical researches in physics are purely logical deduction, such as the Big Bang. Pure physical theory research needs to be combined with philosophy, and some concepts can be better expressed in philosophical language. Modern physics is a science that can use the cross language of physics and philosophy.

The physical mechanism presented by nature is complex. Pure physical research experiments can easily enter the situation of blind people feeling elephants. It is difficult for us to solve the mystery of material structure by a high-energy particle collider. Mathematics is only a logical tool. It is wonderful when the logical relationship in mathematics is consistent with the objective laws. A top physicist should have a brain of philosophical thinking and a complete and materialistic world outlook.

Philosophy is a method to study the universality and particularity of things. It is the research on the essence and development law of the world. It is a discussion about phenomenon and essence. The logical thinking of philosophy can help us look at the problem as a whole, improve the visual height of looking at the problem, and make the outline clear. Therefore, the functions of modern physics and modern philosophy are getting closer and closer. The main difference between them is that one is qualitative as a whole and the other is quantitative analysis in detail.

Matter has different forms. Let's discuss the world we can see first. The visible world is the sun, moon and stars. The universe of the sun, moon and stars is also called the visible universe. The visual universe is a part of the whole universe. Now the question is: how is this part of the visual universe composed. Of course, the universe has many forms of matter that we can't see.

§ 4·1 The main body of the world "the action of matter"

What we are discussing now is actually a question about the nature of the world. What is the nature of the world? This is not only a philosophical problem, but also a physical problem. At the philosophical level, there is a debate between idealism and materialism. In physics, it is the ultimate goal pursued by scientists. However, with the development of science, the understanding of this problem has become more and more confused, especially the emergence of quantum mechanics, which makes this problem close to theology. In fact, the problem lies in the completeness of quantum theory.

What is the nature of the world? Although we know that everything is composed of molecules, atoms, neutrons, protons, electrons, quarks and other material structures, these material structures are the form of matter, not the basis of matter. The essence of matter is not the so-called "basic particle" in any sense.

The so-called "elementary particles" that have now been discovered, there seem to be hundreds. In fact, most of these are "material fragments" after particle accelerator acceleration or high-energy particle collision, not real material structures. There are only a few particles such as electrons, protons and neutrons that can exist stably in nature. According to modern physics, these substances can be divided into two categories - fermions and bosons.

Fermions and bosons are two "particles" with completely different properties. In addition to their different spins, fermions satisfy the principle of "Pauli incompatibility", that is, more than two fermions cannot appear in the same "quantum state". For the time being, we will not consider the boson, but analyze this characteristic of fermion.

All matter particles in "elementary particles" are fermions. Fermions are the raw materials of matter, including quarks and leptons. Quarks and leptons have common properties: first, mass, followed by electromagnetic properties. Mass and electromagnetic properties are the basic attributes that all material "particles" must have. It is precisely because fermions have such material properties that they show the principle of "Pauli incompatibility".

For any substance to exist, first of all, it must have quality. Things without quality do not exist. Quality is the basic characteristic of material existence. The quality that human beings can feel is its weight. Weight is the representation of mass in gravitational environment. The essence of weight is gravitation. Gravitation is the most basic attribute of material existence that human senses can feel.

Another characteristic that human beings can observe the existence of matter is the electromagnetic characteristic of matter. The electromagnetic characteristic is that if a substance is not charged, it must be magnetic. The performance of electromagnetic characteristics is the electromagnetic force that we can observe at ordinary times. Any substance must have electrical or magnetic properties in addition to its mass. The essence of mass is gravitation, and the performance of electromagnetic characteristics is electromagnetic force.

Universal gravitation and electromagnetic force represent the two forces that matter exists. These two forces are the macro forces that human beings can feel the existence of matter. There are two other forces that human beings cannot perceive, the weak force and the strong force. The four natural forces represent all the characteristics of material existence. In nature, things can be various, but in the final analysis, they are all four kinds of natural forces. There is nothing in this world except four natural forces.

Why can the colorful world in front of us be reduced to four natural forces? Because the world we can feel is produced through the interaction of four natural forces. The color seen by the eyes is the interaction of electromagnetic forces. Hand and skin can produce touch, in addition to the role of electromagnetic force, there is gravity.

Although we cannot accurately describe the four kinds of natural forces, everything in the world can finally be attributed to the four kinds of natural forces. The four natural forces represent all the functions of matter. We call the four natural forces "the action of matter". It also stipulates that "the action of matter" is the main body in the world.

When we define universal gravitation, electromagnetic force, strong force and weak force as the main body of the world, no matter what form these forces are behind, the rationality of the definition itself will not be destroyed. How do these four natural forces come into being? What is the essence behind these four natural forces?

If we accept the Big Bang theory, the four forces of matter originate from the Big Bang. It has been discussed previously that the Big Bang is the big bang of space. How did the space explosion form the present space? How does matter in space come into being? The theoretical basis for answering these two questions is Einstein's "general relativity" and Maxwell's equations.

"General relativity" describes "gravitational space", which reflects the physical relationship between flat and straight space and mass. Maxwell equations actually describe "electromagnetic space", which reflects the electromagnetic characteristics of vacuum space. That is, "general relativity" and "Maxwell's equations" describe two forces in nature, universal gravitation and electromagnetic force.

Universal gravitation and electromagnetic force reflect the basic attributes of "material field". The assumed "space big bang" model is: the performance of "material field" is a force. The action of this force seems to want to achieve a certain purpose, and the Big Bang is the action process of this force. This force was eliminated after the Big Bang.

In this way, we can regard the big bang as a neutralization annihilation reaction by two "matter fields" with different polarities. The result of neutralization reaction is that the function of "material field" is "satisfied" and the existence of force is eliminated.

Due to the imbalance between the two "matter fields" with different polarities. After neutralization of the two "matter fields" with different polarities, some of the "matter fields" with polarities are not neutralized. In this way, the big bang in space formed three things:

1. Space: the space before the big bang was actually a chaotic world seriously distorted by the "material field". Later, the clear and transparent space was formed by neutralization and annihilation of two "material fields" with different polarities.

2. The "material field" left after the explosion: The big explosion is a neutralization reaction of two "material fields" with different polarities. Due to the imbalance of the two different polarity "material fields", a part of the "material fields" with polarity did not form

neutralization reaction after the Big Bang. The polar material field without neutralization reaction was left after the big explosion. The sun, moon and stars we see, including ourselves, are made up of this part of matter.

3. Energy is generated: two different polarity "material fields" release huge energy in neutralization and annihilation reactions. This energy is converted from the gravitational potential energy of the "material field", which is similar to the energy released when the rubber band is disconnected.

After the big bang, there are some "material fields" with electrical and magnetic properties left. These "material fields" gain energy in the space formed by the big bang. The leptons and quarks of fermions are the manifestations of this "material field" after obtaining energy.

Quantum field theory holds that particles or waves in the universe are excited states of corresponding quantum fields, and each particle has its own corresponding field. In the microstructure, the "material field" is represented by discontinuous "quantum states", so the "material field" can be called "quantum field". Due to the "quantization" of the "material field", the formed microstructure also has "quantization".

Only the excited state of "quantum field" can be expressed as the existence of matter. The so-called excited state is the "quantum field" after obtaining energy. Energy enlivens the "quantum field" and becomes an existence. The "material field" is a kind of force action, and the essence of force action is that the "material field" has not obtained satisfy. Because the polarity of the "material field" was not neutralized with the opposite sex in the big bang. When the "material field" can-not carry out positive and antimatter neutralization reaction, the longing for the opposite sex becomes the eternal pursuit of the "material field". The pursuit behavior is expressed as force, and the trend of force shows an extreme: the space tends to be infinitely small, and the quality tends to be infinitely large. This extreme behavior of the "material field" can only be balanced with energy.

Failure to neutralize with the opposite sex has become a lifelong regret of the "material field". This makes the "material field" always in the unsatisfied state and always in the action state of force. This unsatisfied state of "material field" extends to all aspects of the world, including human consciousness. Because of unsatisfied, the world has just had pursuit, innovation and development. The variety of the world comes from the dissatisfaction of the "material field" itself, and satisfaction has become the eternal pursuit of the "material field".

Unsatisfied of "material field" is manifested in its physical properties, such as mass, charge, color charge, magnetic moment, etc. These attributes can be summarized into four natural forces: gravity, electromagnetic force, strong force and weak force.

The four natural forces are the characteristics of the material when it is "unsatisfied", representing the existence of the material. The four natural forces are also the only way for us to perceive the existence of the world. The universe formed by the "action of matter" is also called the visible universe.

The action of matter includes four kinds of natural forces. Force has magnitude and direction. We can use a vector to describe this physical quantity. The "material field" in nature always depends on "energy". The magnitude and direction of the force presented by the "material field" are always changing. At this time, we need to use a "wave function" to describe this "the action of matter". "Wave function" describes the process of "the action of matter" changing with time in its natural state. This shows that "the action of matter" is a wave under normal conditions.

This is a very important conclusion. It not only gives the physical connotation of "wave function", but also concludes that the natural state of all microscopic "particles" is a wave. That is to say, all microscopic "particles" are waves under normal conditions.

The action of matter is very magical. It not only formed stellar materials, but also evolved life and even human beings. The essence of life is also a material behavior. In the process of life, there are various material functions. People's consciousness is also full of material effects. The human body produces pleasure under the action of dopamine, leading life to struggle. Pain makes people seek advantages and avoid disadvantages. The action of matter is so wonderful. It starts from material dissatisfaction, develops gradually from small to large, and develops from intangible to tangible. From micro structure to galaxy cluster. From lifeless rocks to conscious humans. It makes every dark corner a colorful world. The action of matter is omnipotent. The action of matter can be so unscrupulous, because matter has energy. Energy gives material life and vitality. In this world, in addition to the main body, the action of matter, there is another main body, namely, the energy of motion, called "the motion of matter".

§ 4·2 The main body of the world "the movement of matter"

There are two forms of motion of matter. One is simple displacement motion. The movement of particles from point a to point b is called motion. Another form is the transformation of one movement form into another. For example, the kinetic energy of the motion of particle a is transferred to particle b, and the motion of particle a becomes the motion of particle b.

Simple displacement motion is a symbol of motion energy and a measure of material motion ability. This energy is a conserved quantity. In classical physics, energy is a measure of the mutual transformation of the motion forms of different substances. It represents the ability of the physical system to do work. Obviously, there are two concepts of energy here. One is to express the measurement of material motion, and the other is to express the ability of motion form transformation.

In cosmology, energy is the ability of all material movements. It includes not only simple displacement movements, but also the mutual transformation of different material movement forms. Energy includes not only the motion of non- inertial system, but also all the motion of inertial system. The original form of energy is simple displacement movement, and the advanced form of energy is the transformation between material movement forms. Both forms can be called "material movement".

We cannot know the specific mechanism of energy generation in the big bang, but after energy generation, it becomes a "subject of existence". The reason why energy or "movement of matter" is defined as "subject of existence" is because energy is the other half of the world. After the big bang, the total amount of energy or "movement of matter" will neither increase nor decrease. The movement form of matter can be changed and transferred in different levels of material structure, but the total "energy conservation" or "momentum conservation". Next, we can use a Newton pendulum to illustrate the essence of energy. See Figure 4.1.

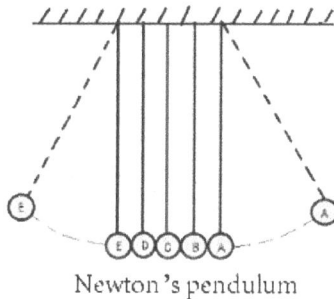

Newton's pendulum

Fig. 4.1

In Figure 4.1, when iron Ball A falls, it can transfer a "thing" to iron Ball E through iron Balls B, C and D, this "thing" is called kinetic energy. If iron Ball A has certain kinetic energy, this kinetic energy will neither be generated nor disappear out of thin air. The kinetic energy of iron Ball A can be transferred to iron Ball E through iron Balls B, C and D, and iron Ball E can also be transferred to iron Ball A through D, C and B. If the iron Ball does not lose energy in the process of transmitting energy, this kinetic energy will always exist.

This shows that the motion of matter is called kinetic energy, momentum, potential energy and energy. It is also the main body in the world. It can transfer between different particle points, and can also be converted into different forms of energy, but it does not grow out of nothing or disappear out of nothing.

Energy is also the main body, which also has a hierarchical structure. In the microstructure, energy also shows the quantum characteristics of one drop, one drop and discontinuity. A share of energy is called energy quantum, and Planck constant describes the size of this energy.

Energy has both the smallest structural unit and the unity of the whole. The unity of the whole is manifested in the indivisibility of energy, such as the energy of a photon. Indivisibility is the rigidity of energy. Rigidity makes energy behave as a particle at the moment of action. For example, light behaves as particles when refracted and reflected. The rigidity of energy means that energy is always a whole in the process of transmission and transformation. Either none or yes is displayed. There is only 0 or 1 status during detection. This property of energy makes energy behave as particles when it acts.

Although energy is also the main body, it is a dependent parasitic main body. When the host to which the energy is attached collides, the energy may be transferred. The collision of energy transfer is called "non elastic collision", otherwise it is "elastic collision". Only in the process of collision can energy transfer take place.

When two hosts collide with each other, the collision will become the action of energy. The main body of the collision is energy. When the main body of energy acts, the host of energy parasitism will also show the rigidity of particles. Since the natural state of all microscopic "particles" is a wave, this function of energy makes the microstructure have wave particle duality.

Energy forms include potential energy, mechanical energy, chemical energy, electric energy, magnetic energy, thermal energy, nuclear energy and so on. The essence of these energy forms is the kinetic energy of material movement. Heat energy is the irregular movement of atoms, molecules and other material particles. Electric energy is the directional flow of electrons. Potential energy is transformed from the kinetic energy of an object's motion. When the kinetic energy of an object's motion is bound by gravity, the original kinetic energy becomes the potential energy in the gravitational field. The essence of the transformation of kinetic energy into potential energy is that the moving particle jumps from low-energy orbit to high-energy orbit. See Figure 4.2.

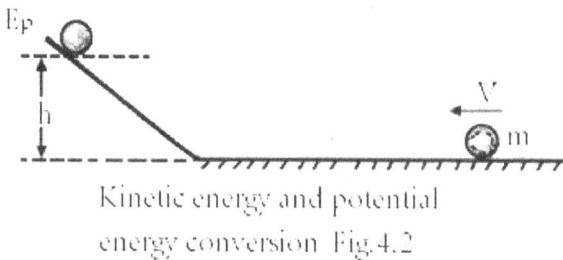

Kinetic energy and potential
energy conversion Fig.4.2

When a moving particle is bound by the "material action" force, the original kinetic energy of the particle becomes potential energy. Chemical energy is the kinetic energy of electrons bound (absorbed or released) by electromagnetic force. Nuclear energy is the kinetic energy of neutrons and protons bound by strong force. The essence of each energy form is kinetic energy, which belongs to the "movement of matter". The "movement of matter" is bound by the "action of matter" and becomes other energy forms. The original form of any energy is "the movement of matter".

In addition to being a conserved quantity, energy or "motion of matter" has another property that is its dependency property.

Although energy or "motion of matter" exists as the main body, it is a parasitic main body and cannot exist alone. Motion needs to be attached by matter, and energy must have a carrier. Without matter or carrier, motion cannot exist. The "attachment attribute" of energy or "movement of matter" reflects the relationship between matter and movement, and the direct expression of this relationship is organism and life. Organisms must have life, otherwise they will die, and life must depend on organisms, otherwise life cannot exist. Organism and life are the two main bodies. They are not the same thing, but they cannot be separated. Energy and matter are interdependent.

Another property of energy or "movement of matter" is its spatial nature. Energy has spatial attribute. Space is the stage of energy, and energy must occupy geometric space. When the form of energy changes from kinetic energy to potential energy, or energy is transferred to a "substance" and becomes the internal energy of the substance, the spatial attribute of energy endows the "substance". When this "matter" obtains this energy, the matter has a spatial structure. Once the "matter" loses energy, it means that the "matter" loses the geometric space it occupies.

"The action of matter" can bind the energy of motion and turn the energy into the "internal energy" of matter. The greater the "internal energy" contained in a substance, the stronger its vitality. When matter loses energy, it also loses its living space. The loss of living space means that matter will become Infinitesimal. Quality, energy and space constitute the "life characteristics" of matter.

Boson is a particle with an integer spin quantum number. In fact, they are the main body of energy, representing the movement of matter or the existence of energy. Due to the attachment property of motion or energy, motion or energy must be attached to matter. Boson is such energy particles that exist with the help of matter. For example, photons are energy particles whose energy exists with the help of electrical and magnetic properties. Although neutrinos are classified as fermions, the main component of neutrinos is energy.

Fermion and boson, one represents "the action of matter" and the other represents "the movement of matter", both of which are the main body of the world. Because of the dependency property and spatial characteristics, the two main bodies cannot exist alone, and they need to depend on each other. Fermions, which represent "the action of matter", contain energy. The boson, which represents the energy of motion, also contains the action of matter. Fermion and boson represent the existence of the two main bodies of matter and motion respectively.

§ 4·3 Strong "equivalence principle"

There are two main bodies in the world: "the action of matter" and "the motion of matter". Now we will discuss the relationship between the two main bodies. The weak "equivalence principle" mentioned earlier is that "gravitational mass" is equal to "inertial mass". The "gravitational mass" of a particle is a measure of the interaction between the particle and the surrounding "gravitational space";

"inertial mass" is a measure of the interaction between the particle and the surrounding "flat and straight space". The physical meaning of the weak "equivalence principle" is: in the "flat and straight space", the mechanical effect in the reference system of accelerated motion is equivalent to the mechanical effect of the reference system in the gravitational field; that is, the acceleration motion of the particle is equivalent to the gravitational action on the particle.

There are two forms of motion of matter, one is the relative motion of simple displacement property, and the other is the transformation of motion state. Under the inertial "reference frame", the motion generated by force is acceleration motion. Acceleration is the change of motion state. Since the world itself is in motion, the motion we are talking about is the transformation of the state of motion, not the transformation from immobility to motion. The previous weak "equivalence principle" actually reflects the equivalent relationship between "gravitational field" and "acceleration".

The acceleration in the weak "equivalence principle" describes the motion of macro particles, and gravity is the force within the framework of universal gravity. We already know that gravitation is only a force in the function of matter, and the movement of macro particles is only a form of movement in the movement of matter. "the action of matter" and "the movement of matter" have a broader physical connotation.

What is "the action of matter" and what is "the movement of matter"? These concepts are now very clear. They are the two main bodies in the world. Can we expand the "action of matter" from universal gravitation to all natural forces; extending the motion of matter from the motion of macroscopic particles to all physical behaviors; if "the action of matter" is still equivalent to "the motion of matter", it means that the "equivalence principle" is applicable to all physical behaviors and is a universal law.

"The action of matter" and "the movement of matter" are philosophical concepts. The physical quantities expressed by them have a common physical essence. For example, "motion of matter", which can describe the displacement motion of a particle, or an energy form, such as potential energy. We use a philosophical concept to discuss physical problems, because the "equivalence principle" does not strictly limit the physical objects it stipulates. It is more accurate to explain the "equivalence principle" with the equivalence of "the action of matter" and "the movement of matter", which expresses the universality of the "equivalence principle". In this way, the "equivalence principle" is applicable not only to the macro world, but also to the micro field; it applies not only to gravitation, but also to electromagnetic force or other forces.

We limit the "action of matter" to the scope of universal gravitation, and the "motion of matter" to the acceleration of macro particles. This "equivalence" is called the weak "equivalence principle". Expand the "action of matter" to four natural forces and the "movement of matter" to all physical changes; if the "action of matter" and "movement of matter" are still completely equivalent at this time, it is called the strong "equivalence principle".

Strong "equivalence principle" is a broader concept, which includes all the action and movements of matter. It is the equivalence between the two the main body of the world we talked about earlier. If the strong "equivalence principle" is established, its physical significance is very profound, which shows that the "equivalence principle" is a universal law followed by matter.

The weak "equivalence principle" reflects that in the framework of universal gravitation, the gravitation of a particle is equivalent to the acceleration motion of the particle. In fact, this is the physical relationship expressed by Newton's second law F=ma . In electromagnetic interaction, the most intuitive movement is the directional flow of electrons - current. Current I represent the movement of matter, and electromotive force E or voltage represents "the action of matter". Figure 4.3 shows the equivalent relationship between current and voltage:

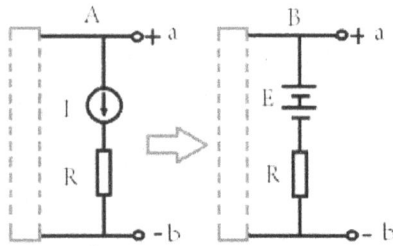

Current and voltage equivalent
Fig. 4.3

If there is current flowing through the resistance R in circuit A, the "current source" I in circuit A can be replaced by an "electromotive force" E and become the form of circuit B. In circuit B, electromotive force E=IR. From ports a and b of the circuit, circuit A is completely equivalent to circuit B. This is the equivalence of "motion of matter" and "action of matter" in electromagnetic interaction.

In the electromagnetic interaction, the most vivid expression of the "equivalence principle" is the motor and generator. The motor reflects that the voltage can become mechanical motion, and the generator expresses that the mechanical motion can also become voltage. "The action of matter" and "the movement of matter" can completely convert each other in electromagnetic interaction. At the macro level of the action of gravitational force and electromagnetic force, the "equivalence principle" is fully established.

Among the four kinds of natural forces, weak force and strong force are the manifestations of "material action" in the micro field. The motion of microscopic particles is a behavior involving weak force and strong force. At the macro and micro levels of material structure, the "equivalence principle" has different physical mechanisms. At the structural level of macro matter, "the action of matter" is completely equivalent to "the movement of matter"; in microstructure, "the action of matter" and "the movement of matter" are expressed as a unity or community.

What is the unity or community of "the action of matter" and "the movement of matter"? This involves Einstein's most famous formula, the mass energy equation $E=mc^2$. Let's first look at $F=ma$ of universal gravitation, the relationship between voltage and current in electromagnetic force, $U=RI$, and the mass energy equation, $E=mc^2$. If c^2 is replaced by $v, E=mv$. The three have exactly the same mathematical expression. The mathematical formula in macro structure expresses the equivalence between "the action of matter" and "the movement of matter". It shows that the "movement of matter" is caused by the "action of matter". In the microstructure, the "movement of matter" is equivalent to a share of energy E. what does this mean?

The micro particles belong to non-neutral structure. In addition to its basic mass, a microscopic particle also exhibits electrical or magnetic properties. The forces in the microstructure include weak force and strong force. After studying various relationships, we found that the essence of weak force and strong force is the force presented by the mass and electromagnetic characteristics of matter. The "equivalence principle" here is shown as: "the action of matter" equals to "the movement of matter"; "The action of matter" and "the movement of matter" are the same individual. Weak force and strong force in microstructure have corresponding energy respectively. The mass energy equation expresses the energy corresponding to the mass of matter.

In the mass energy equation, if the speed of light takes the natural unit $C=1$, then $E=m$, which means that mass equals energy. If the mass is m_0, then the energy carried by m_0 is E_0. m_0 refers to the "gravitational mass" of the microstructure, and E_0 refers to the energy equal to the mass m_0.

In the microstructure, in addition to the energy corresponding to the mass, there is also the energy corresponding to the electromagnetic force. Electromagnetic force is actually two kinds of force, magnetic force and electric force. Magnetic force has corresponding energy, and electric power also has corresponding energy. The essence of strong force is magnetic force, and the energy corresponding to magnetic force is nuclear energy. The energy corresponding to electricity is the kinetic energy of the electron.

In the microstructure, "the action of matter" and "the movement of matter" are equivalent, or the same individual. This is a matter with a mass of m, and its existence carries such energy as E. The individual being is the combination of mass and energy. If "the action of matter" is not "gravitational mass", but the charge of electrons. The charge action of electrons is also the "action of matter", which also has corresponding energy. If the effect of electric charge is e_k, the energy corresponding to e_k is E_k, and the total energy carried by this electron is $E=E_0+E_k$. Since m_0 is a "gravitational mass", the energy E_0 corresponding to m_0 is the structural energy inside the electron. The energy corresponding to e_k is E_k, and E_k is the kinetic energy of the overall motion of the electron.

The above is the expression of "equivalence principle" in microstructure. Can these theories be established? Why does m_0 have the corresponding energy E_0 in the microstructure? What is the basis of these theories? Why does the "equivalence principle" have different forms from the microstructure at the macro level? The theoretical basis for answering this question is "balance principle".

Now we know why the speed of light does not change, it is because of the "Doppler redshift". The reason why the strong "equivalence principle" is tenable is because of the "balance principle". "Balance principle" is the basic law followed by material structure. The "balance principle" and the "equivalent principle" complement each other and constitute a physical mechanism. Mass energy formula is the most intuitive mathematical form of "equivalence principle" and "balance principle" in microstructure.

§ 4·4 equilibrium principle

In the previous section, we discussed the "equivalence principle". The reason why the "principle of equivalence" can be established is that the principle that substances follow in the structure is "balance". The visible universe is a world connected by forces. When the space in the universe is a physical entity, each individual in the space is under the action of some force. In addition, there will be interactions between individuals. Therefore, the force balance of interaction is the basis for the stable existence of all things.

Windmill is the characteristic symbol of Holland and the image ambassador of material movement. The windmill stands still when there is no wind, but moves when there is wind. A tall windmill under the blue sky interprets the universal principle of the existence of everything in the world. When the windmill is stationary, the blades are in balance under the action of gravity. When the windmill rotates, the windmill is in the balance of force under the action of wind. Whether stationary or rotating, the windmill is always in balance under the action of force.

All things in the world can be a windmill or an electron. The existence of things is always in a balanced state under the action of force. If there is an asymmetric unbalanced force in the system structure, there must be another force to keep the system structure in balance.

Let's first analyze the force in the system structure within the scope of universal gravitation. Under the action of universal gravitation, a system or a particle, which can be an electron, a molecule or a windmill, must be a force balance structure. When it is in a relatively static and stable state, its total external force must be equal to zero, that is:

$$\sum_{i=1}^{n} F_i = 0 \qquad 4-1$$

In the formula, F_i represents the vector component of a certain force in the system structure. If the total resultant force in a system or system is not zero, for example, the blade of the windmill is subjected to the force of the wind, this excess force will inevitably lead to the change of relevant physical quantities, for example, the blade rotates with the wind force. At this time, the structure of the force in this system is:

$$\sum_{i=1}^{n} F_i = m_i a \qquad 4-2$$

In the formula, m_i is the mass of a physical quantity and a is the acceleration. After the wind blade rotates, the windmill is in a new equilibrium state. According to the "equivalence principle", the motion of particle m_i is equivalent to a force F, F= ma. At this time, the mechanical structure of the whole system is:

$$\sum_{i=1}^{n} F_i - m_i a = 0 \qquad 4-3$$

The windmill system increases the blade movement. The reverse force generated by the blade movement makes the whole force structure in the system in a new equilibrium state, and the combined external force of the system is still equal to zero. At this time, the structural balance of windmill force is dynamic balance, and the motion is the motion in balance.

In the framework of universal gravitation, the combined external force of a closed physical system must be equal to zero. This is the dynamics characteristic of a physical system without any gravitational field. It shows that under the condition of vacuum space, any existing individual is in a relaxed and force free state. For a closed physical system to be in a stress-free state, the system must be a force balance structure. Only when the stressed structure of the system is a balanced system can the system not interact with the outside world.

In the environment of gravitational field, the combined external force of a closed physical system is equal to zero, and the physical system must have a certain acceleration movement to release the gravitational effect of the gravitational field. When the physical system moves with a certain acceleration so that the combined external force of the system is equal to zero, the dynamic characteristics of the physical system are the same as the vacuum space environment, and the system is in a state of no external force.

The physical system in vacuum environment, like that in gravitational environment, must be a force equilibrium structure. Balanced structure is the Dynamics feature of all macro structures. For the electric and magnetic forces in a system, the system should first pursue the balance of internal electric and magnetic forces. The charge force is the balance of positive and negative charges; Magnetic force is the closure of S and N poles of magnetic force lines. There are negative end potentials in nature, and there must be corresponding positive end potentials. There is no magnetic monopole in the natural state. If a single terminal potential or magnetic monopole exists, other forces must be used to maintain the balance.

In the electromagnetic interaction, if the positive and negative charges cannot be balanced or the magnetic lines of force cannot be closed smoothly, the system will also use the repulsion generated by motion to maintain the balance. In the atomic structure, when the nucleus is positively charged and the electrons are negatively charged, the electrons will move around the nucleus. Electrons use the repulsion generated by motion to counteract the gravity of positive charges. "Movement of matter" can not only resist universal gravitation, but also counteract electromagnetic force.

Electromagnetic force is the force that forms the structure of matter. When the repulsive force of "material movement" is balanced with electricity and magnetic force, the structure of matter has spatial attributes at the same time. "Material movement" needs to occupy space, and space is the attribute of "movement". When the motion characteristics are used in the material structure to maintain the balance of the system, this spatial attribute is given to the material structure.

Whether it is gravitation or electromagnetic force, the force action tends to be extreme, either infinitesimal or infinite. It is impossible to build the world by relying on "the action of matter" alone. For this extreme trend of matter, another force is needed to balance it, and balance is the basis for the existence of things.

The above is the balance principle of universal gravitation and electromagnetic force in the macro structure of matter. Gravitation and electromagnetic force have their own balance mechanisms. For the same force, the balance mechanism in the macro structure is different from that in the micro structure. To understand the difference between macro and micro balance mechanisms, we must first understand what is macro structure and what is micro structure. What is the difference between the two?

The substance depends on electromagnetic interaction to form its structure. The difference between macro and micro is the electromagnetic characteristics of the substance after it forms its structure. At this time, we do not need to consider the role of gravity: if the structure formed by electromagnetic interaction is the balance of electromagnetic force, that is, the "neutral structure", the material structure belongs to the macro structure. If the structure formed by electromagnetic interaction is a non-equilibrium body of electromagnetic force, that is, "non neutral structure", the material structure belongs to the microstructure.

The "neutral structure" is the equilibrium body of electromagnetic force. After the gravitational action is balanced, the particle is always in a static state in an environment where the absolute temperature is zero. Atom is a "neutral" material particle except for gravitational effect, which belongs to the balance of electromagnetic force. Therefore, atoms are the smallest particles of matter that we can observe at rest.

Material particles smaller than atoms belong to the microstructure. The microstructure belongs to "non neutral structure", which is a non-equilibrium body of electromagnetic force. In addition to gravitational action, the "non neutral structure" also presents electrical or magnetic characteristics. In order to maintain the balance of the structure, various forces in the microstructure always correspond to the energy of equivalent motion, which maintains the balance of the structure.

Therefore, the microstructure does not need external force to push, and the structure itself is in motion. The microstructure belongs to the community of "movement of matter" and "action of matter".

The "balance principle" is used to the macro or micro structure of all substances. The difference between macro and micro is "balanced structure". Without considering the quality problem, if the system as a whole is a "balanced structure", its motion attribute belongs to the macro. If the system as a whole is a "non-equilibrium structure", its motion attribute is microscopic. The macro needs to be driven by external forces to produce movement, and the microstructure itself is in motion. The same electronic motion. The movement of electrons in the atomic orbit is microscopic, and the directional movement of electrons - current is macroscopic. The macroscopic current needs external voltage to produce movement.

Why not consider the quality issue when making qualitative macro and micro differences? The balance of mass action has its own unique mechanism, and different levels of material structure have different balance ways. The mass balance relationship in the microstructure is determined by the mass energy formula. In the macrostructure of condensed matter larger than atoms, the role of mass is gravity. The balance of gravity is completed by the structural mechanics system, such as the desktop support force. In the structure of galaxies, the balance mechanism of mass effect is guaranteed by the complete equality of "gravitational mass" and "inertial mass".

The mass energy formula expresses the "equivalence principle" and "balance principle" in the microstructure: in the microstructure, "mass" and "energy" are completely equal, which means that two subjects with opposite properties construct a balance system. The weak "equivalence principle" reflects the "equivalence principle" and "balance principle" in the structure of galaxies: in the structure of galaxies, "gravitational mass" and "inertial mass" are completely equal, representing that the mass action and motion energy are completely equal.

Balance is not only the basic principle of the operation of the universe, but also the principle of material structure. The rotation of the windmill and the movement of the universe are the same balance mechanism, and the atomic structure and celestial structure follow the same balance principle. The existence of any kind of thing is a "balanced structure". Because of this, the "equivalence principle" at the macro level shows that "the movement of matter" is equivalent to "the action of matter". In the microstructure, the "equivalence principle" shows that "the movement of matter" and "the action of matter" are combined into one.

"The action of matter" and "the movement of matter" are the two main bodies of the world. The "equilibrium structure" formed by the two main bodies has very profound physical connotation. The "action of matter" has the property of one-way convergence, and its limit is infinitesimal. The "motion of matter" or energy is just the opposite in nature. It is a multi-directional and divergent thing with infinite spatial orientation. Only when these two opposite natures of the main body are combined and reach the balance of the two forces, can the structure formed by matter exist stably.

PART FIVE

Material forms in the universe

Earlier, we discussed the universe from three levels. The first level is that through the red shift of the spectrum, we realize that space is also a physical entity. From the fact that space is a physical entity and space is expanding, it is believed that the main body of the Big Bang is not matter, but space. The Big Bang is an explosion of space. The Big Bang is the expansion of space. It not only solves the problem that matter does not grow out of nothing in the Big Bang, but also upholds such major principles as "conservation of mass" and "conservation of energy". The theory of space inflation is also consistent with the expansion of the universe in reality.

The second level is about the nature of the world. Put forward the two main bodies of the world: "the action of matter" and "the motion of matter". The four kinds of natural forces are called "the action of matter", and all energy is summed up as "the movement of matter". It also sums up some main characteristics of the two main bodies. "The action of matter" force is a vector force with magnitude and direction. Since it is always dependent on energy, the magnitude and direction of force are always changing, which needs to be described by a wave function. Therefore, the original form of matter is a wave. "The motion of matter" has spatial properties, dependency properties, rigidity, quantum properties, etc. which provide a theoretical basis for us to solve the mystery of material structure.

The third level is "equivalence principle" and "balance principle". The "equivalence principle" and "balance principle" are the basic principles of the world composed of matter. When we know the space, matter and the structure principle of matter, and the theoretical basis of these three aspects, we can discuss the structure and motion law of the universe.

The space before the Big Bang was actually a chaotic world seriously distorted by the "material field". This view comes from the cognition of the "black hole". Later, the clear and transparent space was formed by neutralization and annihilation of two "matter fields" with different polarities. The essence of the Big Bang in space is the neutralization reaction of two different polarity "matter fields". This assumption comes from the understanding of universal gravitation and electromagnetic force. Due to the imbalance of the two different polarity "matter fields", after the Big Bang, a part of the "matter fields" with polarity remained. Because this part of the "material field" has polarity, the polarity allows the "material field" to be connected, which forms the material structure. This material structure is the sun, moon and stars we see.

§ 5·1 "Equivalent equilibrium state" and visual universe

The first thing we want to talk about is the visible universe. The visible universe is the sun, moon and stars that we can see. From the previous discussion, we can see that the visible universe is composed of the remaining "matter field" with polarity in the Big Bang of cosmic space. Although the Big Bang is a hypothesis, all the substances constituting the visible universe have an obvious feature, which is that the substances constituting the visible universe have electromagnetic properties. Electromagnetic characteristics are the basic attributes of matter, while mass is not the basic attribute of matter. The electromagnetic characteristics of matter originate from the polarity of "matter field".

The electromagnetic property of a substance refers to that if the structure of the substance is not charged, it must be magnetic. Therefore, substances with electromagnetic characteristics can be connected between substances through electromagnetic force to form structures and form the sun, moon and stars. Matter has electromagnetic characteristics, so that matter can emit light by radiating electromagnetic waves to be seen by us.

The electromagnetic force presented by the electromagnetic characteristics is just a kind of binding force that connects substances. To form a real material structure, matter needs another thing, which is "energy". The material structure formed by pure electromagnetic force does not have spatial characteristics. Space is the attribute of energy, and electromagnetic force needs to be combined with energy to make the formed material organization have spatial characteristics. For example, the atom occupies a certain space volume because the electrons in the atomic structure contain energy. The energy of electrons endows the atomic structure with spatial properties.

The big explosion in space not only formed an empty and transparent space, but also produced a lot of energy. After the big explosion, there were some "matter fields" with polarities left. These "material fields" not only gained "energy" in the Big Bang, but also had space. Space is also a kind of "resource". Space and energy make the "matter field" that was originally in a high-density entanglement state diffuse loosely into the space formed by the Big Bang.

When the space expands further, the temperature will gradually decrease with the expansion of the space. These "material fields", which obtain energy and have polarity, become a relatively free and independent material structure when the gravitational action and energy repulsion between them just keep a balance. This material structure is called "Fermion".

Fermions are basic material particles with electromagnetic properties and certain energy. It is the "quantized" form of "matter field" after obtaining energy. Since this material structure is a "balanced structure", it can be seen from the previous "equivalence principle" that this balance is formed through the role of "equivalence principle", so the state of matter in this structure is called "equivalent equilibrium state".

The so-called "equivalent equilibrium state" of matter is the equilibrium structure composed of "the action of matter" and energy. In the equilibrium structure of this kind of matter, there are at least two kinds of force effects, mass effect and electromagnetic effect; there are at least two kinds of balance, the balance between mass and "energy", and the balance between electromagnetic action and "energy".

After "energy" and "mass" are balanced, energy becomes the internal energy of material structure. The "mass" of matter is m_0 in the mass energy formula, and the energy is the energy E_0 corresponding to m_0. If it is an electron, E_0 is the structural energy inside the electron. If the total energy of the electron is E, then $E - E_0 = E_k$, and E_k is the kinetic energy of the overall motion of the electron. In addition to the equivalent balance constructed by mass m_0 and energy E_0, an electron also has the equivalent balance constructed by electronic charge e_k and kinetic energy E_k.

A microstructure is a balanced superposition state constructed by multiple actions, which at least includes gravity and electromagnetic force. Gravity and electromagnetic force are both "the action of matter", but in the microstructure, only the balance of electromagnetic force needs to be considered. The "equivalent equilibrium" corresponding to gravity has been standardized by the mass energy formula in the process of forming fermions.

The interactions in the microstructure are mainly electric and magnetic. The value of the energy corresponding to electricity and magnetism is related to the spatial position of the "particle", so the kinetic energy E_k is a state function.

Electricity and magnetic force are the forces that form the

structure of matter. Force has magnitude and direction, which is a vector physical quantity. Because the electromagnetic force in the microstructure needs to be balanced with the energy, it is always in motion. The magnitude and direction of the force are always changing. Therefore, the electromagnetic force can use frequency υ. Wavelength λ To describe it, this is the wave function. The wave function describes the continuous distribution of electric field or magnetic field in space, which shows that the microstructure itself is a kind of wave.

The energy of a free particle is a definite value, and its motion state can be described by a wave function. The energy of a particle in a bound state, such as an electron in a potential well, is related to the motion state of the particle. The wave function describing the motion state of this electron must satisfy the Schrodinger equation. Schrodinger equation describes the continuous distribution of electric or magnetic fields in a microscopic system and the corresponding energy relationship. This correspondence reflects the balance between electricity or magnetism and energy.

The natural state of microstructure belongs to the action state of electricity and magnetic force, such as electrons, neutrons and protons in atoms. Therefore, their form can only be a wave. When matter is in wave form, energy maintains the "equivalent balance" of wave form, and the energy is an implicit function of matter wave. Any kind of microstructure is a dual main body. When one side dominates, the other side exists as a hidden variable. When a collision occurs, the "equivalent equilibrium state" of the wave may collapse into a particle to absorb or release energy. When the microstructure absorbs or releases energy, it is the energy that plays a role. At this time, the microstructure will completely show the rigidity of "particles".

Since the microstructure of matter is an "equivalent equilibrium state" constructed by "the action of matter" and "energy", the microstructure of matter has the following behavior characteristics.

1.Wave particle duality: when the main body of "the action of matter" plays a role, it is shown as wave; when the main body of "energy" works, it behaves as a rigid particle. Whether a microstructure is represented as a wave or a particle is determined by the main body in action. Since the microstructure contains two main bodies, the microstructure has wave particle duality.

2.Quantum property: The motion distribution of the charge or magnetic moment in the microstructure is a state function. When its motion state changes, the corresponding energy E_k will also change, but the energy absorbed or released can only be a small share, a small share, and a "discontinuous" amount. If a physical quantity has the smallest indivisible basic unit, the structure of this physical quantity belongs to the "quantized" structure, and the smallest structural unit is called quantum. Energy has the smallest structural unit and indivisible overall characteristics. The "action of matter" corresponding to energy is also a discontinuous physical quantity. Therefore, the behavior of microstructure has "quantum" characteristics.

3.Uncertainty principle: microstructure is the continuous distribution of electric charges or magnetic moments in space. Its essence is a wave. The energy carried by the wave has a minimum resolution unit h. If the physical parameters of a particle are used to accurately describe it, if the displacement error is Δx. The momentum error is $\Delta P_{\lambda\chi}$. The product of two kinds of errors is greater than or equal to $\hbar/2$, namely: $\Delta\chi^* \Delta P_{\lambda\chi} \geqslant \hbar/2$. This shows that if the microstructure is treated as a particle for quantitative analysis, its position and momentum cannot be accurately described at the same time.

4.Observation effect: The microstructure is an "equivalent equilibrium state" with wave particle duality. Observe and measure a microstructure, which may be a wave or a particle. Whether the observation result is wave or particle depends on the means of observation and measurement. Any observation and measurement will destroy the original balance. If the measurement is an energy type collision, energy transfer will occur during the collision, and then the microstructure will collapse into a particle. It is a probability event that we can detect the particle in a certain area.

5.Planck constant: The microstructure may show the characteristics of waves, and may also show the rigidity of particles. The energy of a particle can be expressed by displacement $\Delta\chi$, momentum ΔP these physical parameters to describe. For the energy carried by the material wave, $P=h/\lambda$ can be used, $E=h\nu$ give. Here h is Planck constant, which is a very important physical parameter in microstructure. The energy and momentum carried by material waves are related to Planck constant h. Planck constant h is not only an important physical constant, but also an energy unit. Microstructure is a kind of wave in the natural state, and wave is the carrier of energy. The energy of one cycle of material wave oscillation is h, and the unit is j*s. Planck constant h also reflects the smallest energy structure in quantum. H stands for the equivalent energy of "the action of matter" within a cycle. A period is the smallest integer unit of the change of "material action", and the corresponding energy is also the smallest energy unit. The size of energy can only be an integral multiple of it.

In quantum mechanics, one misinterpreted is the superposition principle of "states". Microstructure is an "equivalent equilibrium state", which is similar to the middle equilibrium point of seesaw. It can change into "0" state or "1" state. We can't use the superposition state of "0" and "1" to describe this "equivalent equilibrium state", because the superposition state of microstructure has its specific physical connotation.

The "state" of a microstructure must be an equilibrium system, which can form a new structure with other "states", and the new structure must also be an "equilibrium state". The superposition principle of "state" is actually the balance principle. A microstructure is an "equilibrium state", which can be superimposed with other "equilibrium states". After superimposing, it must still be an "equilibrium state".

Quantum entangled state: "state 1" is an "equilibrium state" and "state 2" is also an "equilibrium state". If "state 1" can form a "superposition state" with "state 2", this "superposition state" must also be an "equilibrium state". The reverse is not necessarily true. If a "superposition equilibrium state" composed of "state 1" and "state 2" cannot be disassembled into two relative independent "equilibrium states", it is said that "state 1" and "state 2" are entangled, and the two sides are "entangled states" to each other. "Entangled state" is also a kind of "quantum state".

Pauli's "incompatibility" principle explains the balance requirements of the "superposition state". In a system composed of fermions, two or more particles cannot be in the same state. Fermion is an "equivalent equilibrium state" constructed by the "action of matter". When two particles with the same properties form a "superposition state", in order to maintain the overall balance of the system, the two particles are always in a reverse action state.

Microstructure is an "equivalent equilibrium state" constructed by "the action of matter" and "the movement of matter". The material structure larger than the atom belongs to the macro structure. The macro structure of matter is neutral. Without considering the effect of gravity, the neutral structure itself is an equilibrium body. Each atom can not only exist alone, but also rely on electromagnetic force to form a new equilibrium structure.

Whether macro or micro, matter forms its structure through electricity and magnetism. In the macro material structure, new material structure can be formed through the balance of electricity and magnetic force. The material structure formed by the balance of electricity and magnetic force is called "chemical bond". Metal bonds rely on free electrons to form an electrical balance; ionic bonds are balanced by positive and negative ions depending on each other; covalent bonds maintain equilibrium by sharing electrons. In this way, matter forms various things and forms the world around us.

For a large mass celestial body, in addition to the balance of electromagnetic force, the gravitational effect caused by the overall mass also needs the equivalent repulsive force to maintain the balance.

The mass in the mass energy equation is the mass at the micro level. The macroscopic mass of a massive celestial body and the gravitational effect formed are the main internal forces of matter, which also need corresponding repulsive forces to maintain balance. The repulsive force inside the material structure mainly includes the thermal movement of the material particles, the repulsive force between the same sex charges, and the electron degeneracy pressure. When the repulsion inside a material structure cannot resist its own gravity, the whole material structure will collapse and explode; collapse from one "equivalent equilibrium state" to another "equivalent equilibrium state" with lower energy level. The supernova explosion is the collapse of the "equivalent equilibrium state" of the star.

Astronomical observations have found that all kinds of stellar explosions are actually the collapse of stars. The gravity of a large mass celestial body needs equal motion energy to construct an equivalent balance. If the energy contained in a star cannot maintain an "equivalent equilibrium state", it will cause the star to explode. The result of the explosion is that the star collapses from one "equivalent equilibrium state" to another "equivalent equilibrium state" with lower energy level (see "non-equilibrium state" and black hole).

When the level of material structure rises to the level of galaxy structure, the gravity caused by mass becomes the main force between stars in the structure of galaxy. In the structure of galaxies, "Equivalent equilibrium" is the balance between gravity and inertial force. The physical mechanism for maintaining the structural balance of a galaxy is that the "gravitational mass" is completely equal to the "inertial mass".

In the mass balance mechanism, galaxies (macro) and micro have similar balance structures, which are maintained by motion characteristics, that is, equivalent balance. Originally, only gravity and repulsion are required to achieve equilibrium, but the balance mechanism in the microstructure is that "mass" and "energy" are completely equal. The equilibrium structure of galaxies is that "gravitational mass" and "inertial mass" are completely equal.

The mass is equal to the energy, and the "gravitational mass" is equal to the "inertial mass". Their balance principles are similar. They are all a balance mechanism with negative feedback properties. In the microstructure: the greater the mass, the greater the energy carried by the mass. In the structure of galaxies: the greater the "gravitational mass", the greater the "inertial mass". This kind of negative feedback can make the balanced structure keep balanced and stable even if there is a large range of force fluctuations.

Hydrogen atom is the simplest "equivalent equilibrium" structure in the world, and helium atom is also an "equivalent equilibrium" structure. The difference between them is the energy. Due to the role of the balance mechanism, the balance mechanism will build an "equivalent equilibrium state" according to the amount of energy the system has. When the energy possessed by the system cannot maintain an "equivalent equilibrium state", the equilibrium mechanism will reconstruct a new "equivalent equilibrium state". There are more than 100 kinds of atomic structures in the world. Their components are the same. They are electrons, protons and neutrons. The difference is that the system contains different energy.

The balance of galaxies is that the "gravitational mass" is exactly equal to the "inertial mass". When the "gravitational mass" of a massive object is very large, the "inertial mass" against gravity will also increase, which can maintain the structural stability of galaxies. The structure of the galaxy will not cause the galaxy to come apart due to the fluctuation of gravity in space or the change of motion speed.

Another advantage of "gravitational mass" being completely equal to "inertial mass" is that every celestial body can be in a state of equilibrium between gravitational and repulsive forces, regardless of its mass. The balance makes every celestial body seem to be in a free state without force. The weightlessness experienced by the astronauts in the space station is such a state of no force. A secret of nature is hidden here: on the one hand, there are universal connections and interactions between substances, making the whole universe a whole; on the other hand, every individual is in a state of freedom without force. Only when every individual is in a free state of being free from force, can this individual get the opportunity of free development. With such a living environment, the earth can evolve and develop intelligent life - human beings in 5 billion years.

The world we see or perceive is an "equivalent equilibrium state" composed of "the action of matter" and the energy of motion. "The action of matter", under the coordination of energy, jointly constitutes atoms, molecules, aggregated substances and even life forms. A life form is an advanced form developed by "the action of matter". In the advanced form of life, "the action of matter", energy also evolves into life of organisms. Organism is the advanced form of "the action of matter", and life is the advanced form of energy.

In the advanced life of human beings, "the action of matter" generates various hormones through human tissues such as glands and organs, or forms special physical effects to stimulate human organs, thus forming human consciousness. Human consciousness has become a super high form of "the action of matter". People's desire, ideal, hope, belief, etc. are the expression of "the action of matter". Here, energy becomes a kind of vitality, a kind of power and a kind of spirit. Although human society is a very complex physical system, it is still composed of "the action of matter" and "the movement of matter".

Whether it is a single celled organism or a conscious and intelligent human being, life has a common feature, which is the yearning for the opposite sex. In addition to the need to reproduce offspring, sex is the original trace of the world's material properties. Why does matter work? Because the world is composed of the remaining polarity "matter field" in the Big Bang. Why does the "material field" have polarity? Because the polarity of matter was not "satisfied" in the Big Bang. Therefore, the desire for the opposite sex has become the eternal pursuit of "material action". Obviously, this feature of "material action" also dominates the behavior of life.

§ 5·2 "Nonequilibrium state " and black holes

The "equivalent equilibrium state" of matter is the visible universe we can observe, which is the sun, moon, stars and everything around us. Is there any other form of matter? have. What is discussed below is another form of matter - the "non-equilibrium state" of matter.

The "equivalent equilibrium state" of matter is an existing form supported by the energy of "material movement". "Energy" is the vitality of all things. Without energy, nothing can survive. Our life is like this, and so is material life.

A complete individual of matter is called an atom, and when it was first born it was called hydrogen. A proton plus an electron, like a human newborn baby, is simple and full of energy. The chapter of material life is to burn itself, release light and heat, and thus reveal its own existence.

Early stars were mainly composed of hydrogen, which continuously fused hydrogen into helium through the catalysis of carbon, nitrogen and oxygen cycle. The way stars live is also the main way of material survival. They all emit light and heat through nuclear fusion reaction, which is the existence value of matter. This way of life of matter is determined by the characteristics of matter containing energy.

The life of stars is mainly the fusion process of hydrogen nuclei, which is also the main star sequence stage of stars. The life of matter is determined by the energy contained in the matter. The ratio of energy to mass of hydrogen atom is the highest. It becomes helium atom through nuclear fusion and releases the corresponding energy. The fusion of hydrogen atoms into helium atoms means that matter has grown one year old. Helium atoms are also full of energy. It is still the puberty of matter. With age, matter will gradually lose energy.

Although the life of matter is accompanied by the evolution of stars, the life of matter is different from that of stars. Some matter has to undergo several evolutions of stars. Matter can regain energy in stellar explosions. Energy rejuvenates heavy elements into light elements, and then forms stars to emit light and heat. But as a whole, matter always develops from light elements to heavy elements. Slowly exhaust all its energy in the light and heat, and complete its life.

The evolution of matter is essentially a metamorphosis of "equivalent equilibrium state". The transformation of matter is generally from high-energy state to low-energy state. Each "equivalent equilibrium state" is an equilibrium structure supported by a certain amount of energy. The balance of the structure has a dynamic range. If the balance of the structure fluctuates within the dynamic limit, the structure is stable. If the fluctuation of equilibrium exceeds the limit, the original "equivalent equilibrium state" will collapse. In the collapse of the "equivalent equilibrium state", the substance can absorb energy and transform to the high energy level. For example, in a supernova explosion, the ejected material can obtain higher energy. It can also be the material that releases energy and converts to a lower energy level, such as the center of a supernova explosion. However, in general, the transformation of "equivalent equilibrium state" is always from high energy level to low energy level.

The process of material life is the process of energy loss. When the energy contained in its own structure cannot support an "equivalent equilibrium state", this "equilibrium state" will collapse and the material structure will change to another "equivalent equilibrium state" of low energy level. For example, a hydrogen atom is converted into a helium atom.

The conversion between different "equivalent equilibrium states" of matter is an absorption or release of energy. The absorption or release of energy will be reflected in the space volume ratio occupied by the material structure. Energy has spatial property. The more internal energy a system contains, the more space it occupies. For example, the hydrogen element contains the most energy, and the space occupied by hydrogen with the same mass is much larger than that of iron element.

In the material hierarchy, energy can be divided into internal structural energy and kinetic energy of overall motion. Internal structural energy is the inherent energy of the system, and kinetic energy is the ability of the system to move as a whole. In the material hierarchy, the kinetic energy of the next order is the internal energy of the previous order. Therefore, the larger the internal energy contained in a system or system, the larger the space occupied by the system. The size of the space occupied by a system is a sign of how much energy a substance contains. Energy has spatial properties. The larger the space occupied by matter, the more internal energy it contains.

In the process of evolution, matter will release and consume its internal energy. When a system releases and consumes part of its internal energy, and the original equilibrium state can no longer be maintained, the equilibrium state will collapse. Such collapse is the transition from the high energy level equilibrium state to the low energy level equilibrium state. In addition to releasing excess energy, the space occupied by the system stability will also be reduced. If all the internal energy of a system is released, the space occupied by the system will tend to be infinitely small.

Space is the attribute of energy and the symbol of material vitality. During the evolution of matter, after the fusion of atoms into iron nuclei, matter will become inactive. Since iron cannot undergo nuclear fusion and release energy, if the gravitational force of the star is greater than the repulsive force, the star will further contract under its gravitational force. For small mass objects, the "degeneracy pressure" of electrons can prevent the further collapse of the gravitational force of stars, and stars form white dwarfs. If the "degeneracy pressure" of electrons cannot prevent the collapse caused by the gravitational force of the star, the electrons will be pressed into the protons, and the star will become a neutron star.

Although white dwarfs and neutron stars are highly dense stellar bodies, their matter also has a structure, which belongs to the "equivalent equilibrium state" of matter, and the material structure also contains energy. For the celestial body with larger mass, the gravitational force will cause the celestial body to further contract, and the contraction will release the energy of the system, and the release of energy will further contract the system. This is a positive feedback process, which will cause a system to release all the energy in the structure. When these celestial bodies lose their last bit of energy, they will become a "black hole" in space.

The matter that loses any energy also does not have any space structure. At this time, the matter is a "matter field" entangled by gravity. At this time, the space volume occupied by the "material field" will tend to be infinitely small; the mass per unit volume will tend to infinity. This state of the "material field" is called the "non-equilibrium state" of matter. "Non-equilibrium state" is a state in which matter loses all its energy.

The existence of matter must be an "equivalent equilibrium state". When a "equivalent equilibrium state" cannot maintain equilibrium, the material structure will collapse. After a "equivalent equilibrium state" collapses, the system will enter the "equivalent equilibrium state" of low energy level. When the energy in the system cannot maintain any equivalent equilibrium, the system will enter the "non-equilibrium state".

Nuclear fusion and celestial explosion belong to the collapse of "equivalent equilibrium state". When atoms fuse from hydrogen to iron, each nuclear reaction will lose part of the energy and collapse from the high energy level to the low energy level. For small mass stars, when the hydrogen in the core of the star is burned out, gravity will cause the core of the star to shrink, and the core contraction will release excess energy. The released energy is transferred from the center to the periphery, resulting in the expansion of the outer shell material of the star. The explosion of "red giant" is actually the process of star collapse. Supernova explosion is also the collapse of stars. The collapse of each "equivalent equilibrium state" is because the energy can no longer maintain the original equilibrium, and a new "equivalent equilibrium state" of low energy level needs to be established.

The system will enter a new "equivalent equilibrium state" after collapse. Collapse is the contraction of the space occupied by the original structure of the material and the release of excess energy. When the energy contained in the material structure is too low to maintain any balance, the material enters the "non-equilibrium state". Since the "non-equilibrium state" of matter is a mass of matter without any energy, when a massive star collapses into a "black hole", it will release all its energy at once.

We can define a "black hole" in this way: if the energy contained in an equivalent equilibrium system can no longer maintain the equivalent equilibrium of the whole system. And the system can-not continue to maintain balance by changing the structure, motion state and reducing the space volume. At this time, the gravitational action in the system is completely greater than the repulsion of energy. Gravity will further collapse the whole structure. The collapsed system will crowd out and release all the energy in the system. In this way, the space occupied by the system will shrink to an infinitesimal point. This point is the "black hole".

Because energy has the attribute of attachment, there are two ways for the collapse of stars and celestial bodies to release energy: one is to radiate energy through the electromagnetic characteristics of space, and the other is to throw energy out through matter. Through the electromagnetic characteristics of space, the radiated energy is the radiated electromagnetic wave, and the radiated electromagnetic wave spectrum ranges from microwave to gamma ray. By means of releasing energy from matter, in addition to emitting various high-energy particles, there is also a jet of material carrying high energy.

Radiating electromagnetic waves and emitting various high-energy particles are normal phenomena in the thermonuclear reaction of stars. When a star or celestial body collapses into a "black hole", the main way to release energy is to eject high-energy material (material jet). Small mass stars, such as the sun, evolve into red giants. The final collapse of the core and expansion of the shell are also a way to throw energy outward. Supernova explosion is the same, the core shrinks and the outer shell material carries energy and is thrown out.

"Black hole" is a state of matter after matter has lost all its motion energy. It is black because there is no energy in the matter that constitutes the celestial body. "Black hole" is also a celestial body, which may devour other stars. When the energy of the engulfed star can establish a new balance with the gravitational action of matter including the "black hole", the "black hole" may disappear. Obviously, this situation is only possible when the "black hole" is a small mass object.

When the "black hole" is a massive celestial body and the energy of the engulfed star cannot balance the gravitational action of matter including the "black hole", the "black hole" will throw out all the energy of the engulfed star. The "quasars" in the "active galaxies" observed by astronomy belong to this astronomical phenomenon.

"Quasar" is a kind of celestial body that looks like a star. In fact, it is an older galaxy, which is the evolution of galaxies in the early universe. In the early universe, the space was not as vast as it is now. The density of stars in the space was high and the mass of galaxies was large. Therefore, the evolution of "quasars" was characterized by that on the one hand, they quickly became "black holes" and at the same time, they constantly created new primitive nebulae.

"Quasar" consists of "active galactic nucleus", "accumulation disc" and "jet flow". The so-called "active galactic nucleus" is actually a super "black hole" swallowing the surrounding stars. "Accretion Disc" is composed of stars swallowed by "black holes". When the "black hole" devours the star, it will spit out all the energy contained in the star. This forms the high-energy jet flow of "quasar". Jet flow from a "black hole" is a nebula that later forms stars in space. After the "black hole" spits out the material with high energy, the remaining material is the "non-equilibrium state" of the material without any energy, which becomes a part of the "black hole".

Jet flow has huge energy because the "black hole" spits out all the energy of a star in an instant after swallowing it. This is why the energy released during a jet flow activity may be greater than the energy released by a galaxy in its lifetime. After the star is swallowed by the "black hole", the original "equivalent equilibrium state" collapses instantaneously and directly becomes a "non-equilibrium state". This collapse process is more violent than the process of releasing energy when the "equivalent equilibrium state" degenerates.

The "escape speed" of a "black hole" reaches the speed of light, which means that the "escape speed" of matter cannot escape from the black hole even if it reaches the speed of light within the horizon of the "black hole". The "black hole" does not emit light because the "black hole" has no energy, not because the "light" cannot compete with gravity and does not emit light. If a "black hole" can emit light, light can escape from the "black hole". Light is an electromagnetic wave and is not affected by gravity, that is, the "gravitational mass" of light is zero.

The "black hole" radiates energy or ejects high-energy material flow outward, which is the case when the "black hole" devours the stellar material. Under normal circumstances, "black holes" will also receive electromagnetic radiation and high-energy particles from space. Since the "black hole" does not contain energy and has no material structure, the physical mechanism of how to deal with this part of energy received by the "black hole" is still unclear. The so-called "Hawking radiation" may be related to this. The electromagnetic radiation and high-energy particles received by a "black hole" under normal circumstances must be different from ordinary matter. It involves the evolution of "black holes". "Black hole" is the state of matter after death. The rebirth of "black hole" reflects the cyclic evolution of matter. The cyclic evolution of matter is the epitome of the evolution of the whole universe.

The universe has a history of 13.8 billion years. The early universe was not so vast, and matter existed in the structure of super large "galaxy clusters". Its mass is billions of times that of the sun. The evolution period of massive galaxies is only a few million years. At present, the centers of super large "galaxy clusters" are all super "black holes", which control the operation of the universe. The universe we can see is basically composed of hydrogen and helium, but after the death of matter, it is a "black hole". They are "nonequilibrium" states of matter. The "non-equilibrium state" is also a universal state of material existence.

§5·3 "Absolute equilibrium state" and vacuum

The visible universe is the material world formed after the energy is obtained by the "material field" with polarity. The evidence is that any material constituting the world has electromagnetic characteristics, and the electromagnetic characteristics are the polarity of the "material field". Similarly, we now believe that vacuum space is formed by neutralization and annihilation of two opposite polarity "matter fields". The evidence is that vacuum space also has electromagnetic characteristics, but the force of space electromagnetic action is in absolute equilibrium, which makes the space body hide its electromagnetic essence.

We can understand the polarity of "material field" through the characteristics of electric field and magnetic field. The polarity of the "material field" is a strong pursuit of the "opposite sex", and it is a manifestation when the "satisfaction" is not obtained. When two "matter fields" with opposite polarity collide, neutralization effect occurs. This neutralization effect is the Big Bang. The Big Bang eliminated the polarity of the "material field", eliminated the existence of any force, and did not show any external characteristics. It was an "absolute equilibrium state" of electricity and magnetic force. When two opposite polarity "matter fields" are neutralized, a vacuum space is formed.

We can also understand the electromagnetic effect of vacuum space through electric field and magnetic field phenomena. If there is a charge Q in the vacuum, the charge Q will destroy the balance of the original charge in the vacuum space, and the vacuum space will generate an electric field around the charge Q. The reaction of vacuum space to charge Q in space can be described by Maxwell equation.

Maxwell equation $\oint E^*ds = \dfrac{Q}{\varepsilon_0}$ describes the charge characteristics of vacuum space. If there is a charge Q in the vacuum, the charge Q will destroy the balance of the original charge in the vacuum space, and the vacuum space will generate an electric field around the charge Q. The expression of Maxwell's equation is that the electric field flux passing through the closed surface enclosed by the charge Q is equal to $\dfrac{Q}{\varepsilon_0}$, where ε_0 is the dielectric constant of vacuum. It should be noted here that the electric field around the charge is not carried by the charge, but is the reaction of the electrical and magnetic properties of vacuum space to the charge Q.

An erroneous concept in modern physical theory holds that Maxwell's equation describes the physical properties of electricity and magnetic field, which is actually wrong. We know that there will be an electric field around a point charge in space and a magnetic field around the magnetic pole. However, the electric field and magnetic field are not carried by the charge and magnetic pole itself. If they are inherent in the charge and magnetic pole itself, they cannot be separated from the charge and magnetic pole and transmitted to the distance as electromagnetic wave signals. The electric field around the point charge in space and the magnetic field around the magnetic pole are produced by vacuum space, which is the manifestation of the electromagnetic characteristics of space. Maxwell's equation reflects the electromagnetic characteristics of vacuum space, not the properties of charge or magnetic moment itself. Electric field or magnetic field is only the performance of the physical properties of vacuum space. Without this physical property of vacuum, the electric field itself will not become a magnetic field, and the magnetic field itself will not become an electric field.

Equation $\oiint B^*ds=0$ describes the magnetism of vacuum space. Vacuum space is an equilibrium state under the action of magnetic force, and the magnetic line of force is a closed passive field. The equation states that the integral sum of the magnetic flux passing through the closed surface in space is always equal to zero whether there is a magnetic field in the vacuum or not. If any closed surface in the space has magnetic flux inflow, there must be equal magnetic flux outflow, because the vacuum space is the "absolute equilibrium" of magnetic force.

Equation $\oint_l E^*dl + \iint \frac{\partial B}{\partial t}^* ds = 0$ describes the equilibrium characteristics of the overall electromagnetic properties of vacuum space. If there is an electric field E in the vacuum, make a path integral around the closed curve of the electric field E, which is equal to the rate of change of the magnetic flux B on the integral surface. After adding the two terms, it is equal to zero. This also means that the electric field in the vacuum must have the function of the corresponding magnetic field, and the function of the magnetic field must also have the function of the electric field. The two are equal in size and opposite in function. The vacuum space always maintains the "absolute balance" of electricity and magnetism as a whole.

The previous equation reflects that the change of magnetic field in vacuum can produce electric field. Similarly, the change of electric field can also produce magnetic field. Equation $\oint_l B^*dl = I\mu_0 + \mu_0\varepsilon_0 \iint \frac{\partial B}{\partial t}^* ds$ reflects the current I in a conductor or the changing electric field

B, which will cause a magnetic field in the surrounding space. Vacuum space is the "absolute equilibrium state" of electrical and magnetic characteristics. The change of current and electric-field in the conductor will cause the change of vacuum magnetic field. Vacuum maintains the absolute balance of electrical and magnetic characteristics as a whole.

If there is an open terminal potential at a certain point in the vacuum, and the electromotive force is an alternating electric field. According to the equation, the changing electric field will cause the changing magnetic field in space, and the changing magnetic field will also cause the changing electric field in space. This change of electromagnetic properties in vacuum space is electromagnetic wave. Electromagnetic wave is an energy carried by vacuum space, "electromagnetic space" is the vacuum space disturbed by energy.

Electromagnetic wave is the energy carried by "vacuum field", radio signal is electromagnetic wave, radar signal is also electromagnetic wave, visible light is electromagnetic wave, and laser is also electromagnetic wave, because these are the energy radiated into vacuum space. The energy radiated into the vacuum space in the Big Bang is still there today. It fills the whole universe. This is the cosmic microwave "background radiation".

In addition to the electromagnetic characteristics of vacuum space, another attribute of space is the flat and straight characteristics of space. The "absolute equilibrium" state of matter is not only limited to the balance of electricity and magnetism, but also includes the flatness and straightness of space. Flatness and straightness do not lie in the form of space, but in the characteristics of space. Flatness and straightness represent the desire of the "vacuum field". Being in the state of "absolute balance" is the ultimate goal of the "material field". Absolute balance is the performance when the "material field" gets "satisfaction", and no effort is required for the "material field" in this state. The "vacuum field" of vacuum space will try its best to maintain this equilibrium state of matter.

Cosmic space is a flat and straight space, and the mass matter in the space will destroy the original flat and straight space, making the space curved. The linear propagation of light represents the flatness and straightness of space. We can see the light behind the massive celestial body, not because of the gravitational effect of the celestial body, but because the space around the celestial body is bent. Flat and straight space will bend under the action of mass matter, which can be described by Einstein's field equation. Maxwell equation describes the electrical and magnetic properties of space, and Einstein field equation describes the geometric properties of space. The field equation expresses the relationship between mass matter and space bending. Here, we mainly discuss how the matter in space produces mass and gravity.

The expression of mass in space is gravitation. In a gravitational field, the mass behaves as a vertical downward force. These two forces have the same essence. How does the gravity of an object come into being, that is, how does mass come into being?

Let's do a thought experiment first. If there is no gravitational field deep in the universe, there is only a single celestial body in space.

On this lonely planet, gravity also exists. For the planet itself, gravity means that every tiny part of the sphere is affected by the gravity of the earth's center. According to the nature of force, the gravity of a sphere itself can be regarded as a force that compresses the sphere in the space around the sphere. The space around the sphere compresses the sphere, which is a kind of repulsive force of space and a kind of exclusion of the sphere by space. That is to say, the gravitational force of the sphere itself and the repulsive force of space can be regarded as the same force. The two forces are equal in size, opposite in direction and the same in nature.

If the space around a particle has a force repelling the particle, it is equal to the gravitational force of the particle itself. Similarly, if there is gravity between a sphere and another particle, it is equivalent to treating the particle as a part of the sphere; the gravitational force between the particle and the sphere is equal to the repulsive force generated by space on the particle and the sphere as a whole.

The current view is that any matter has gravity, so it is called universal gravitation. Gravitation is a force described by Newton. Newton's law of universal gravitation can be used to describe the motion of celestial bodies very accurately. Obviously, the law of universal gravitation is correct. How universal gravitation comes into being is the same question as why any matter has mass.

There will be many problems if we regard universal gravitation as a kind of gravitation possessed by matter itself. For example, the problem of force source, the problem of over distance action, the problem of compatibility between micro quality and macro quality, etc. Many problems can be easily solved if we understand gravity as an exclusion force in space.

The big bang in space is a neutralization reaction of two "matter fields" with different polarities. After the neutralization reaction, space is formed. In addition to not presenting any "the action of matter" externally, space also has flat and straight geometric characteristics. We call this state of matter in the universe the "absolute equilibrium state" of matter. The "absolute equilibrium state" of matter not only covers the balance of various forces, but also includes the geometric characteristics of flat and straight space.

The remaining polar material fields that did not participate in the neutralization reaction in the big bang formed a kind of tissue after absorbing energy. The existence of this kind of organization will destroy the flat and straight characteristics of space. The influence of this kind of organization on the flat and straight characteristics of space is called "quality". The size of quality is a measure of the impact of this organization on the horizontal and straight characteristics of space. The greater the mass, the greater the influence on the horizontal and straight characteristics of the space, which will lead to the more significant bending of the space.

All Polarity "matter fields" in space that do not participate in neutralization reaction will affect the flatness and straightness of space. Therefore, they all have mass. We call this organization "mass matter". The size of the mass not only indicates the number of "mass matter", but also indicates its influence on the flat and straight characteristics of space.

For a space with flat and straight characteristics, it will repel "foreign matters" in the space. When there is a "mass substance" in space, the space will try to compress the "mass substance" to minimize its volume. In our view, the force of space compression is more like the contraction force of mass matter itself - gravity.

If the "foreign matter" in the space is composed of two "mass substances", the space will regard the two "mass substances" as a whole, and the space will compress the whole in a larger scope. However, the vector center of the spatial repulsive force will move to the center of mass of the two "mass substances". By analogy, it will be superimposed layer by layer and expanded step by step until it reaches the whole universe. The space composed of many "mass substances" is called "gravitational space". Vacuum space will compress the scope of "gravitational space", and the repulsion of vacuum space seems to be the gravitational effect between various particles. The repulsive force in vacuum space is completely equivalent to the mutual attraction of celestial bodies in "gravitational space". According to the nature of the force, the repulsive force in the vacuum space is the same as the gravitational force in size, opposite in direction and the same in nature. The two forces can be regarded as the same force.

Because the mass of the earth will affect the flatness and straightness of space, space will compress the earth with the center of gravity of the earth as the center of the space repulsion vector. When the moon is a part of the earth's mass, space will compress the earth and the moon as a whole with their centers of mass as the center of the space repulsion vector. Because the center of the spatial repulsive force vector is the center of mass of the mass matter. When there are multiple centers of mass in a celestial system, there are also multiple centers of repulsion vectors in space. Moreover, the repulsive force of space to each center of mass exists simultaneously.

The earth has its center of gravity. When the earth and the moon are as a whole, they have a new center of mass composed of two particles. When the moon revolves around the earth, the common center of mass of the earth and the moon will also drift. The drift of the common center of mass of the earth and the moon means the drift of the space repulsion force vector center, or the drift of the gravity center. On Earth, this drift causes the tides of the sea; in the structure of Mercury's motion around the sun, this drift causes the additional precession of Mercury.

Gravity is not the attraction of mass matter, but the repulsive force of space itself. Since there is no real gravity between substances, "gravitational waves" certainly do not exist. If the change of gravitational action can form a "gravitational wave", then the motion of a massive celestial body will also produce a "gravitational wave". Facts have proved that the "gravitational wave" formed by the gravitational changes of celestial bodies has not been detected so far. Will the repulsive force of space itself produce an effect similar to "gravitational wave"?

Vacuum space is the "absolute balance" state of the "material field". The laws governing the movement of celestial bodies in space are "conservation of energy" and "structural balance". The so-called "gravitational wave" is actually the vibration of space itself. This is only when two massive objects collide in space and release part of the energy to space when they collide; because of the "conservation of energy", this part of energy can only be transmitted by the jitter of the

matter. In this way, the force of particle motion can resist the space repulsion force of particle in the direction of motion. Therefore, the moving particle can continue to move forward along the original direction of motion.

In the direction of motion, the resistance of space to the particle's advance is proportional to the particle's motion speed. When a particle is stationary, the spatial repulsion force on the particle is determined by the mass of the particle. The greater the mass, the greater the spatial repulsion force. When a particle moves, the space resistance of the particle in the forward direction is in direct proportion to the speed of the particle movement. The greater the movement speed, the greater the space resistance. The original motion speed of the particle determines the current motion speed of the particle. The current motion speed of the particle is the balance point between the original motion speed of the particle and the space resistance.

The velocity of a particle moving along a straight line is determined by the original velocity of the particle. The original motion speed of a particle is a forward force. When the original motion speed force of a particle is equal to the repulsive force in space, the particle will keep the original motion speed constant and move forward in a straight line.

When the particle moves forward along a straight line, the motion will cause the space with bending deformation to move with the motion of the particle, as if the space around the particle also moves with the particle in a uniform straight line. The relationship between particles and space is still locked by the same spatial repulsion.

Because the space repulsion force in the direction of particle motion moves backward with the particle motion. When the force of forward motion of the particle and the spatial repulsion force in the direction of motion of the particle reach equilibrium, the equilibrium point is the speed of uniform motion of the particle. Therefore, the space repulsive force in the direction of particle motion does not really prevent the particle from moving forward, but only regulates the uniform motion of the particle.

Since vacuum space is the "absolute equilibrium state" of matter, the energy of moving particles will not be transferred or lost in the interaction with vacuum space. The force exerted by a moving particle on the space of the moving direction is a constant force that will never be exhausted, and the size of the force depends on the original motion speed of the particle. Therefore, the space repulsion force has created another function of the inertia of "mass matter" - "the mover is always moving".

Inertia is that the particle is no longer just static under the "vacuum field" criterion. If the original particle is stationary, the particle remains stationary. If the original particle is in motion, it will continue to move in a uniform straight line. This characteristic of a particle can be described by an "inertial mass".

From the above discussion, the "inertial mass" of a particle comes from the role of space repulsive force. Conversely, the particle has an "inertial mass", which can prove that the universal gravitation is not the gravitation of matter, but the repulsive force of space. The repulsive force of space comes from the flat and straight characteristics of space. Any matter in space has mass, because the mass matter will affect the flat and straight characteristics of space.

Vacuum space is a whole woven by "vacuum field", which connects everything in the universe and contains everything in the universe. The "vacuum field" integrates the whole universe into a whole, and this whole is still an "absolute equilibrium state". The "absolute equilibrium state" makes all the related physical quantities in the universe and the movement of any part of them affect the overall balance. Therefore, the "absolute equilibrium state" of vacuum space has rich and profound physical connotation.

The "absolute equilibrium state" of vacuum space, on the one hand, represents the balance of electromagnetic force of space body, on the other hand, represents the flatness and straightness of space. These two characteristics of space will regulate all physical behaviors in space. The specification of "vacuum field" on matter involves all aspects of matter, making matter have mass is only one aspect, which also includes the way of existence of matter, symmetry of geometric shape, symmetry of force vector direction, etc.

The "absolute equilibrium state" of vacuum space also shows that space has a real physical property of "zero tolerance". The vacuum space can make a physical response of 100%, to "the action of matter" and "the movement of matter". The efficiency of transmission and conversion of "motion energy" in vacuum space is equal to 100%, and its energy consumption is equal to zero. This is why the "gravitational wave" caused by the merger of massive objects can travel billions of light-years to us. The first ray of sunshine from the Big Bang is still wandering in space.

PART SIX

The "Vacuum Field" Standard of Physical Behavior

If vacuum space is also regarded as a physical entity, then the sun, moon and stars are celestial bodies moving in a physical entity. Fish live in water, and all their activities will be limited by water. " The action of matter " and "the movement of matter" are activities in the vacuum space, and all their behaviors will be regulated by the "vacuum field". "The action of matter" and "the movement of matter" represent all physical behaviors in the universe. If all physical behaviors in the universe will be regulated by the "vacuum field", then the vacuum space is the manager of the world.

The theory of the three states of matter can explain many physical phenomena, such as the complete equality of mass and energy, and the complete equality of gravitational mass and inertial mass. But these do not solve the fundamental problem, and do not solve the problem of why. When we read the physical structure of the vacuum space, and when we think about whether the vacuum space has a normative effect on all physical behaviors in the universe, we suddenly have a feeling of sudden openness. It seems that we have walked out of the physical labyrinth of the secret of the universe, and we have obtained the true meaning of the truth. Isn't vacuum space the "God" that people have been searching for?

Vacuum space is a physical entity, which contains everything, regulates everything, and plays the role of "God". What is the principle of physics? What are the laws of physics? Why is the movement of matter regular? The standard of "vacuum field" for physical behavior is the principle of material operation, the physical laws, and the physical laws. Any mysterious physical phenomenon, you can find the answer in the "vacuum field" specification.

Any social form of human beings must have corresponding social systems, laws and regulations to maintain, including culture and morality, in order to operate normally. The same is true in nature. The sun rises on time every day, the four seasons change in spring, summer, autumn and winter, and the sun, moon and stars operate in an orderly manner. All these also need to be standardized. The whole nature needs a manager, which is similar to the role of "God". This "manager" is the "vacuum field" of the vacuum space.

The universe is a whole woven by the "material field". Space contains all the "effects of matter", including all the "movement of matter", and strictly regulates every physical behavior in space. But the gauge action of "vacuum field" is different from other physical functions. The specification of "vacuum field" is only a rule, and it maintains an order. Other actions, such as gravity, can be a continuous force. The normative action of "vacuum field" is similar to the legal system of society, which only plays the role of regulating the behavior of each individual, and its purpose is to maintain the orderly operation of society. When a physical behavior conforms to the specification of "vacuum field", the physical behavior will not be affected. When a physical behavior violates the specification of "vacuum field", the physical entity will be restricted by the force of "vacuum field". The orderly operation of the universe is due to the normative effect of the "vacuum field".

In microstructure, why are energy and mass completely equal, and why are "gravitational mass" and "inertial mass" completely equal? These are the results of the "vacuum field" specification. Why space is full of electromagnetic waves, why energy is conserved, what is weak interaction, what is strong interaction, and so on. All physical laws are the result of the "vacuum field" specification. The normative function of "vacuum field" is the "social system" of cosmic space, which is the legal provisions. All physical behaviors in space must abide by this system. The normative action of "vacuum field" ensures the orderly operation of the world. Many physical laws in nature come from the norms of "vacuum field".

§ 6·1 The normative effect of vacuum field on "material movement"

Vacuum space is not only a physical entity, but also an "absolute equilibrium state". "Absolute equilibrium" is a physical property of space body. This physical property is different from other physical functions, and does not show a specific function. It is only for the whole space body, showing a delicate physical control function. The "vacuum field" that constitutes the vacuum space is mainly used to maintain the "absolute balance" of the space body. The simple operation of "vacuum field" can make a complicated physical system run orderly in various environments.

The "action of matter" and "movement of matter" are the two main bodies in space, and also the main factors affecting the spatial balance. In order to maintain the balance of the space body, the "vacuum field" must first norm the behavior of the two main bodies. Due to the objective existence of the two subjects, the means of "vacuum field" norm: for "action of matter", it is to narrow its action range as far as possible; For the "motion of matter", it is to maintain the momentum of the original motion constant. As long as the momentum of the original motion is constant, the original balance will not change.

The first thing that "vacuum field" needs to be normed is "the movement of matter". The "motion of matter" determines that the "absolute equilibrium state" is a dynamic system. For a dynamic system, what the "vacuum field" can do is to keep the dynamic system in balance in motion. To keep the state of matter always in balance in motion, it is necessary to make the "momentum conservation" of motion. As long as the "momentum conservation" of motion, the original "absolute equilibrium state" can be guaranteed to remain balanced in the dynamic state. Therefore, the nature of material motion can be called: motion in balance, or balance in motion.

Vacuum space is a physical entity, and any "mass matter" in the space will bear the repulsive force of "vacuum field" in the three-dimensional direction. For "flat and straight space", "mass matter" receives the same spatial repulsion force in the three-dimensional direction of space. This repulsive force of space has two functions, one is to make matter produce mass, the other is to fix the particle on the node of space-time, so that "the stationary object is always stationary".

Because the spatial repulsive force on "mass matter" in flat and straight space is equal in three-dimensional space, the resultant external force on a particle in space is equal to zero. That is to say, although any material must bear the repulsive force of flat and straight space, this force only makes the material produce mass and make it have gravity. As long as the particle remains stationary, it will not feel the existence of force except gravity. This is the feeling of weightlessness experienced by astronauts on the space station.

The uniform repulsive force in the three-dimensional direction of "flat and straight space" makes "the stationary object always stationary", and at the same time, makes the moving particle always move in a straight line and at a uniform speed. The "motion" of a particle is equivalent to the "force" in the moving direction of the particle, which can resist the "space resistance" of the particle in the moving direction. When these two forces are equal, it is the speed at which the particle continues to move forward along a straight line. The force in the forward direction of the particle can only maintain the original motion speed of the particle. As long as the motion speed of the particle is greater than the original motion speed, the space resistance will be greater than the force in the forward direction of the particle. Therefore, the repulsive force of space plays the role of regulating the behavior of particles, that is, "momentum conservation".

The repulsive force of space makes the "momentum conservation" of motion. The norms of "vacuum field", like the legal system of human society, stipulate the attributes of momentum and energy ownership. According to the "vacuum field", "only yours is yours". The original static object can only be in a static state, while the original moving object must remain unchanged. Here, the vacuum space uses its flat and straight characteristics to complete three physical functions: first, to make matter have mass. Second, keep the stationary particle still. Third, keep the original velocity of particles in motion unchanged.

The above principle discusses how the "vacuum field" can regulate the physical behavior of particles in the vacuum space without gravitational field, so that the motion of particles conforms to the "conservation of momentum". If there are other celestial bodies in the space, how does the "vacuum field" regulate the physical behavior of particles in the gravitational field to meet the "momentum conservation"?

If m_a is used to represent the "inertial mass" of a particle in vacuum, m_a represents the strength of the particle in vacuum restricted by the "vacuum field". The particle in the gravitational field will also be affected by the gravitational field. The possible action intensity of the particle in the gravitational field is called "gravitational mass", which is expressed in m_g. We now know that "inertial mass" belongs to the specification of the behavior of the particle in vacuum space. When there are other gravitational fields in space, the action mechanism of "vacuum field" on the behavior criterion of the particle is that its "inertial mass" equals to "gravitational mass".

When the "inertial mass" of a particle is completely equal to the "gravitational mass", this makes the "gravitational mass" of a particle completely lose its physical effect. Whether the particle is in the field strength of weak gravitational field or strong gravitational field, the field strength of gravitational field will not destroy the "momentum conservation" of particle motion.

The "gravitational mass" of a particle can be derived from Newton's formula of universal gravitation. In the gravitational field, the gravitational effect of particles with large "gravitational mass" also increases. The gravitational force of a particle in the gravitational field is:

$$F = G\frac{Mm_g}{R^2} \qquad 6—1$$

When a particle receives gravity, it will produce an acceleration a. according to Newton's second law:

$$\frac{GMm_g}{R^2} = m_a a \qquad 6—2$$

Gravitational action is just an "expectation" which represents the purpose of gravity. Because the particle is in both gravitational space and vacuum space. The behavior of particles is not only affected by the gravitational space, but also regulated by the "vacuum field". When the "gravitational mass" m_g causes the particle to accelerate, the "inertial mass" m_a plays the opposite role. Since $m_a = m_g$, m_a and m_g in the formula cancel each other out. The physical meaning of mutual counteraction is that the "vacuum field" constrains the gravitational action of particles. The two mass actions offset each other, and the formula after offset is:

$$\frac{GM}{R^2} = a \qquad 6\text{—}3$$

G in the formula is the gravitational constant, which represents the general environment of gravitational space. M is the mass of the gravitational source, M/R^2 indicates that the field strength in the gravitational space is inversely proportional to the square of the distance, and a is the acceleration.

A particle can gain acceleration in a gravitational field. The magnitude of the acceleration obtained by the particle changes in inverse proportion to the square of the distance from the center of gravity. Because "inertial mass" and "gravitational mass" can cancel each other. When a particle is in a gravitational field, its "gravitational mass" has no contribution to acceleration. The acceleration of a particle is only related to the distribution of the gravitational field in space.

In the same space region where the gravitational field is distributed, regardless of the size of the particle mass, the particles of any mass accelerate in the way of "walking side by side". An iron ball and a feather are in the same gravity field. Although the mass of the iron ball is much greater than that of the feather, the iron ball and the feather can only accelerate in the way of "walking side by side".

Formula 6-3 shows that the acceleration of a particle does not depend on its mass, but on its position in gravitational space. The closer the particle is to the center of gravity, the greater the acceleration obtained. The force causing accelerated motion is not the mutual attraction between particles, but the gradient distribution of the "vacuum field" in space. It reflects that the essence of universal gravitation is not the mutual attraction of "mass matter", but the repulsion of vacuum space.

The essence of the vacuum space is a flat and straight space. When there is "mass matter" in the space, the "mass matter" will squeeze the space and make the "vacuum field" deform. The "vacuum field" will form a three-dimensional gradient distribution of field strength around the "mass material" with the gravity center of the "mass material" as the center and R as the radius. "Mass matter" extrudes space outward, and the vector of the space's reverse force points to the center of gravity. This can be seen as the distribution of "gravitational potential energy" in space. The closer to the gravitational center, the higher the gravitational potential energy, and the higher the acceleration obtained by the particle there. The kinetic energy of the acceleration of a particle is converted from the gravitational potential energy.

In a flat and straight space, the "vacuum field" standardizes the movement of particles to make them have "inertial mass", the essence of which is "momentum conservation". In gravitational space, due to the specification of "vacuum field", $m_a=m_g$. When the "inertial mass" is completely equal to the "gravitational mass", the "gravitational action" cannot make the particle obtain acceleration. The acceleration obtained by a particle in gravitational space is converted from gravitational potential energy. Therefore, in gravitational space, although momentum is not conserved, "energy is conserved".

The above specific examples show that the "vacuum field" uses its own flat and straight characteristics to regulate the movement of matter, so that "the movement of matter" complies with the "conservation of momentum" or "conservation of energy". The essence of momentum and energy is the movement of matter. "Conservation of momentum" and "conservation of energy" finally come down to "conservation of motion of matter". The principle of Newton's pendulum fully explains that in nature, "the motion of matter" is a main body, and the momentum of motion will not be generated out of now here, nor witt it disappear without reason, but can only be transferred to each other.

The movement of matter will neither occur out of thin air nor disappear for no reason, but can only transfer to each other. If the movement of matter cannot change from one form to another, the movement of matter will become another special movement mode. The irregular movement of material particles is called "heat". Thermal motion is the final destination of energy when the momentum of material motion cannot be converted into other forms of energy. In the process of transformation of the movement of matter, there is always some energy that cannot be converted into other energy forms and finally becomes the irregular movement of material particles, which is the law of entropy increase. The law of entropy increase reflects a movement trend of nature, and the "movement of matter" is unidirectional within a certain period.

As for the "motion of matter" and the law of motion, we seem to know what it is at first, but we don't really understand it, let alone why. Now we know that all this is the result of the regulation of the "vacuum field". In order to maintain the original balance of space, the "vacuum field" regulates all physical behaviors in space. The "conservation of momentum" and "conservation of energy" are the results of the "vacuum field" regulation.

Now look back, why is "inertial mass" equal to "gravitational mass". The inertial action of objects comes from the repulsive force of space, and the essence of gravity is also the repulsive force of space. When a particle is in the same space, its inertia and the "gravity" it receives are both repulsive forces of space. The "inertial mass" generated by the repulsive force of a particle in space is of course equal to the "gravitational mass" generated by the repulsive force of a particle in space.

This section discusses the normative effect of "vacuum field" on "motion of matter". The motion here is the macroscopic motion of the object. The normative effect of "vacuum field" on "material movement" is different in macro level and micro level. Because in the microstructure of matter, "the action of matter" and "the movement of matter" are a community, belonging to the same thing. The criterion of "vacuum field" to "matter movement" in microstructure is included in the criterion of "vacuum field" to "matter action". The regulation of "vacuum field" on "material motion" is equal to the regulation of "vacuum field" on "material action".

§ 6.2 The normative effect of vacuum field on "material action"

There are two existential subjects in the visual universe, one is "the movement of matter" and the other is "the action of matter". They represent all physical behaviors in the visual universe. The normative function of "vacuum field" on "material movement" is "energy conservation". What is the normative effect of "vacuum field" on "the action of matter"?

The "function of matter" includes four kinds of natural forces, two at the macro level, gravity and electromagnetic force, and two at the micro level, weak force and strong force. Let's first discuss gravity.

Gravitation is not an attribute of matter. The essence of gravity is the repulsive force of space, or the deformation stress of space. The space of material existence is a flat and straight space, and the role of space presentation is to maintain the flat and straight space. When there is "mass matter" in space, space will compress "mass matter" as much as possible, making the space range occupied by "mass matter" tend to be infinitesimal. Fortunately, "mass matter" contains energy, which has spatial properties. Energy presents a function opposite to space repulsion. When the two opposite forces reach equilibrium, it is the existence form of "mass matter".

Obviously, due to the repulsion of space, matter must have corresponding energy to obtain living space. If matter loses energy, the repulsive force of space will squeeze the action of matter to Infinitesimal. Energy determines the space range of material existence, and also represents the vitality of material. Since the living state of matter is supported by motion energy, according to the "equivalence principle", this state of matter is also called "equivalent equilibrium state". "Equivalent equilibrium state" is the "equilibrium structure" constructed by "the action of matter" and energy.

"Balanced structure" is the basic rule of the existence of matter. There needs to be a proportion range between matter and energy. If the energy possessed by matter is too large, the material organization will fall apart. If the energy is too small, the material structure will collapse and shrink. Therefore, "the action of matter" and "energy of motion" need to have an appropriate proportion, which is the "equilibrium structure" of matter. The normative effect of space "vacuum field" on universal gravitation is finally expressed as "structural equilibrium" of matter.

We can see the basic fact that the visual universe is full of all kinds of celestial bodies. Each celestial body can float freely in space, or be colorful, or radiant, or full of vitality, unless the material "dies". Does gravitation have no effect on them?

Gravitation is the main force in the macro world. If it is regarded as the property of matter, it seems that gravitation wants to gather the whole world under its own command. If the gravitational force is regarded as the repulsive force of space, the repulsive force of space seems to compress all matter together. The real world is not like this. The reason is that the material world is full of energy.

On the one hand, energy makes the world full of sunshine; on the other hand, energy makes materials full of vitality. However, matter and energy need to have an appropriate proportion range, which is the "equilibrium structure" of matter. The "equilibrium structure" of matter is the basis for the free existence of particles in space. It can not only make the whole world run orderly and steadily, but also avoid the extra interaction between particles and space.

When the material structure of a particle is an equilibrium structure: no matter whether the particle is at rest or in motion; whether the particle is in a gravitational field or in a vacuum far from the gravitational field; as long as the interaction between particle and space is not affected by the third party, there is no force confrontation between particle and space. Particle and space, both sides can coexist peacefully.

The gravity that people feel on the ground belongs to the force of the third party. The support function of the ground to people constrains the natural state of people. The gravity that people feel is actually the support force of the ground to people. If a person floats in space, no matter whether there is a gravitational field around him, he will not feel the effect of gravity or inertia. The unstressed state of particles is the natural state of particles under the "vacuum field" specification. This state is also a characteristic of celestial mechanics in space. There is no mutual confrontation, which is the purpose of the "vacuum field" specification. In addition to maintaining the original balance of space, it also symbolizes the principle of peaceful coexistence in the universe.

The norm of "vacuum field" to material movement is "energy conservation", and the norm of "vacuum field" to "material action" is "structural balance". "Structural balance" is also a physical law, which is as important as "energy conservation" and is the basic law that all material structures must abide by. The physical mechanisms involved in "structural balance" are "equivalent principle" and "balance principle".

Electromagnetic force is different from gravity. Electromagnetic force belongs to the polar action of "material field" itself, and electromagnetic characteristics are the attributes of material itself. Electromagnetic force is also a force to promote the balance of material structure; in the structure of matter, positive and negative charges always depend on each other. If there is a negative charge, there is a positive charge. The S and N poles always appear in pairs. However, the balance obtained by electricity and magnetic force is static balance, and static balance cannot build the material world of "equivalent equilibrium state".

In the macro structure, the balance of electromagnetic force does not depend on the regulation of "vacuum field". The macro structure itself is the balance structure under the action of electromagnetic force, but the specification of "vacuum field" establishes the energy structure of matter. Because of the energy in the material structure, the material has vitality. The balance constructed belongs to dynamic balance. For example, electrons in an atom are negatively charged, protons are positively charged, and the whole atom is neutral. The reason why the atom exists is that the electron has a certain amount of energy, which prevents it from colliding with the positron and annihilating. The existence of the visible universe and the orderly operation of the world cannot be separated from the norms of the "vacuum field".

There are four basic forces in nature. In addition to the macroscopic gravitation and electromagnetic force, there are also two kinds of weak and strong forces in the microstructure. When we understand how the "vacuum field" regulates the weak force and the strong force, we should first understand what is the weak force and what is the strong force.

The four natural forces are the basic forces in nature. In order to unify these four forces in theory, Einstein once devoted all his energy in the latter half of his life to studying this subject. Modern physical theory is also a theoretical discussion on this subject. It is impossible to truly understand the "weak force" and "strong force" when we do not understand the normative role of the "vacuum field".

What kind of force is the weak force? The only thing that can be determined now is that the weak force is a basic force in the material structure, and the radiation phenomenon is closely related to this force. The weak force only exists in the material structure of fermions such as electrons and quarks. The main difference between fermions and bosons is that fermions have masses.

Any matter existing in space has mass, celestial bodies have mass, molecules and atoms also have mass, and electrons, quarks, neutrinos and other fermions all have basic mass. We now know that macro mass expresses the influence of matter on space, and gravity is the repulsive force of space, but the mass of micro particles cannot be understood by macro physical effects.

At the micro level, the "material field" of the microstructure will also affect the flatness and straightness of the space, but this effect is the micro effect between the "material fields". To measure the impact of microscopic particles on space, it is also called the mass of microscopic particles, which is m_0 in the mass energy equation. We can also use the "equivalence principle" and "balance principle" to explain why the mass m_0 in the mass energy equation contains the energy E_0 corresponding to its own mass.

Who stipulated the physical relationship of mass energy equation? This is the "vacuum field". The normative effect of "vacuum field" on the universal gravitation in the material structure of fermions such as electrons, quarks and neutrinos is called "weak interaction". "Weak interaction" regulates the balance principle that should be followed in microstructure. That is, the physical relationship expressed by the mass energy equation. If a microstructure cannot meet this balance, radiation will occur. The force regulated in "weak interaction" is universal gravitation, but this universal gravitation is not a macro effect, but the effect of the mass of a micro particle.

We do not know how the "vacuum field" interacts with the mass of microscopic particles in the microstructure, but we can be sure that the physical mechanism of microscopic and macroscopic interaction is different. Analyzing from the physical relationship expressed by the mass energy formula, the microstructure is also a material form maintained by energy, but the corresponding relationship between the mass and energy of the microstructure is determined. If the corresponding equilibrium relationship is destroyed, radioactivity will occur.

The decay of radioactive elements shows that if the original material structure is not a stable "equilibrium structure", the unstable "equilibrium structure" will decay. The decay of particles is the normative action of "vacuum field". Obviously, the weak force is not equal to the weak interaction, and the weak force is the gravitational force of the microstructure. Weak interaction is the gauge action of "vacuum field" on weak force.

This principle is also applicable to strong interactions. If we now think that the weak force is the expression of gravity in the microstructure, then we can deduce what kind of force the strong force is.

The electromagnetic interaction in our concept is a macroscopic phenomenon of electricity and magnetism. It is the electromagnetic force between electrons and nuclei, between atoms, between molecules, and in the hierarchical structure of these substances. The electromagnetic force comes from the electromagnetic characteristics of atomic nuclei, neutrons and protons. The electromagnetic properties of nuclei, neutrons and protons come from the inside of their tissues. However, in the nucleus, neutron and proton, the interaction is not called electromagnetic interaction, but call it strong interaction. The difference between electromagnetic interaction and strong interaction lies first in the different structural levels of matter: the role of electromagnetic force in the micro level is strong interaction, and the role of electromagnetic force in the macro level is electromagnetic interaction. Strong interaction is just because in the microscopic material structure, the action distance is shorter and the force is stronger, so it is also called strong force.

So, is the strong force of strong interaction the same force as the electromagnetic force of electromagnetic interaction?

Nuclei, neutrons and protons are composed of quarks, and the main interaction between quarks is color charge. According to "quantum chromodynamics", quarks have these main characteristics: charge, color charge, spin and mass. The mass belongs to the weak interaction range, the spin belongs to the energy category, and the rest is the charge and the "color charge". Charge is a known thing, and what we don't understand is "color charge".

What is "color charge"? We can learn from the physical properties of neutrons and protons composed of quarks. After comparing the physical properties of neutrons and protons with the main properties of quarks, we can come to the conclusion that the "color charge" of quarks in neutrons and protons is the magnetism of these particles. The color of "color charge" represents the polarity of "magnetism", and "red", "green" and "blue" represent different polarity structures.

If "color charge" is regarded as magnetism, and "color" represents the polarity of "magnetism", then we can fully understand the different "colors" of quarks and the strong interactions between quarks. "Colorless" is the self-cycle closure between magnetic polarities. The same sex repels each other and the opposite sex attracts each other, which is the basic rule of the interaction of "colors".

The "color charge" of quarks comes from the polarity of the

"matter field", and the charge and "color charge" are both sides of the polarity of the "matter field". After the combination of different quarks, if the charge and the color charge are "satisfied" at the same time, it is a neutron; if only the "color charge" is "satisfied", it is a proton. Since the "matter field" left over by the Big Bang itself is an imperfect existence, the neutron structure as a whole still has a certain magnetic moment, which is not a perfect stable structure. The lifetime of neutrons is only 15 minutes. In the proton structure, the "color charge" is completely "satisfied", but the positive charge remains, but this non-equilibrium structure is a very stable structure.

The "color charge" of quarks is the magnetism of quarks. The main contradiction between quarks is the polarity of magnetism. The strong interaction is mainly the interaction between magnetic poles. The action between positive and negative electrons is the action of charge, which is shown as the balance of positive and negative charges. The force between quarks is mainly represented by the mutual closure of magnetic polarity, and the existence of magnetic monopoles is not allowed. Magnetic monopoles cannot exist alone, that is, quarks cannot be separated separately. This phenomenon of quarks is called "quark confinement".Another force existing in quarks is the gauge force of "vacuum field" to regulate "electromagnetic action". Strong interaction is the electromagnetic interaction between quarks under the "vacuum field" gauge. The force that quarks exert on space bodies is the force of electromagnetic properties. Since the hadron composed of quarks is a non "balanced" structure, under the "vacuum field" gauge, the non "balanced" structure also needs energy to maintain balance, and this energy is the "gluon" in the hadron structure.

The essence of "gluon" is an energy particle (boson), and the strong interaction between quarks is achieved by exchanging "gluons". The exchange of "gluons" is the transfer of energy, and the principle can be understood by referring to the principle of "Newton pendulum".

"Gluons" are the energy carried by quarks, and "photons" are the energy carried by "electrons". The transmission of energy must have a carrier. The macroscopic "electronic" energy can be transmitted through the electromagnetic characteristics of space in the electromagnetic interaction. The microscopic energy carried by quarks cannot be transferred through space in strong interactions. Therefore, in the process of transferring energy, quarks must rely on the "matter field" to transfer energy. This makes "gluon" have "color charge".

Electromagnetic interaction includes magnetic interaction and electrical interaction. Electromagnetic characteristics are derived from the polarity of the "material field", and electricity and magnetism are two aspects of the polarity of the "material field"; This theory is based on the known relationship between magnetic field and electric field. Quarks form protons, neutrons and nucleons through magnetic interactions; Electrons form the structure of atoms, molecules and condensed matter through the interaction of charges. Electricity and magnetism undertake different missions in different levels of material structure. The reason why matter can form various structures is the result of the "vacuum field" regulating the "action of matter".

The norm of "vacuum field" to "material action" is "structural equilibrium", and the norm of "vacuum field" to "material motion" is "energy conservation". Because the microstructure belongs to the community of "the action of matter" and "material movement". Therefore, the regulation of "vacuum field" on a microstructure is not only the regulation of "material action", but also the regulation of "material movement". Under the standard of "vacuum field", microscopic particles not only require "structural balance", but also need "energy conservation".

When we straighten out the interrelationship between various physical quantities and understand their internal logic, you will be surprised to find that there are only two forces in the world: gravitational force and electromagnetic force. Gravity is the repulsive force of space, and electromagnetic force is the force of matter itself. The above two forces represent the internal (micro) and external (macro) causes for the formation of the structure of matter.

In addition to the requirement of "structural balance" for material structure, the specification of "material action" by "vacuum field" also includes spatial symmetry. Vacuum space is a flat and straight space, and mass matter will affect the geometric properties of space. Therefore, the specification of "vacuum field" also includes the geometric distribution of matter in space. The geometric distribution of matter in space includes spatial symmetry. For example, an electron cloud is actually a continuous distribution of electric fields in space. In order to maintain the symmetry of the spatial geometry, the s orbit can only be a centrally symmetric spherical structure. The P orbit can only be a double dumbbell symmetric structure centered on the axis.

In addition to regulating the mass distribution in space, the "vacuum field" also has balance requirements for the direction vector of electromagnetic force. The direction vector of electromagnetic force in space must be symmetrical on both sides and balanced up and down. For example, if two electrons in the same orbit are left-handed, the other must be right-handed. Only in this way can the direction vector of the force be kept in balance.

In the microstructure, there are both the balance mechanism of gravity and the balance mechanism of electromagnetic force. The two balance mechanisms in the microstructure are different from those in the macro, and they will be shown in the micro level of the material structure. They will affect all physical behaviors in space in different forms and ways. Here we will use several specific examples to illustrate.

§ 6·3 The essence of light and its physical mechanism

In this section, we will discuss the essence of light. Why discuss the problem of light again? Because light is related to another physical property of space - the electromagnetic property of vacuum space. Light reflects the relationship between space and energy. Light itself is the product of the "vacuum field" regulating the movement of matter. Understanding light can deepen the understanding of space and energy. Matter, space and energy are the three main bodies of the universe, and the relationship between them reflects the basic overview of the operation of the universe. Through the discussion in this section, we can also clarify another problem. The so-called bosons are actually some energy particles.

We can understand the nature of light by comparing neutrinos with photons. In addition to photons, there is also an energy particle called neutrinos in the solar radiation energy. According to theoretical calculation, for every three photons produced in the process of solar nuclear fusion, two neutrinos will be produced at the same time. What is the difference between photons and neutrinos? Photon is the energy radiated in atomic structure, and neutrino is the energy radiated in atomic nuclear structure. Photons are the energy released by electrons during transition, while neutrinos are the energy released by proton fusion. The release of energy by electrons into space is to convert orbital potential energy into kinetic energy and release it into space; proton fusion cannot transfer excess energy to space.

Energy is the movement of matter, and motion must be attached to matter, which is the attachment attribute of energy. Because of the electromagnetic characteristics of vacuum space, electrons can exchange energy with space and transfer kinetic energy of motion to space. The electron can exchange energy with space. In addition to the electromagnetic characteristics of space, there is another reason: electric field is an active field. The magnetic field is a passive field, and the magnetic force lines of the magnetic field are closed in the self-cycle. Maxwell's equation $\oiint B \cdot ds = 0$ gives that the integral sum of the magnetic flux passing through the closed surface in space is always zero, regardless of whether there is a magnetic field in the vacuum or not. The fusion of protons is essentially the polymerization of magnetic fields. The magnetic field can-not release excess energy through space in the process of polymerization. Therefore, the proton can only release energy by losing a small part of its mass during fusion. The energy released is neutrino.

Neutrino is the energy released by the loss of part of the mass of the atomic nucleus. Photon is the energy released by the electrons in the atom through the electromagnetic characteristics of space. It is very important for the world that electrons in the atomic structure can exchange energy with space. The condensed state of any matter in the world is composed of atoms. The atom is the smallest unit of complete material structure. The energy that can be released by electrons in an atom is related to the electronic orbit and its structure. The amount of energy released by electrons into space becomes electromagnetic waves of different frequencies in space. Because of this, we can understand the material structure of light-emitting objects through the spectral lines of light. The interaction between electronics and space not only makes the world full of sunshine, but also makes the world colorful.

The electrons in the atomic structure are all in their own energy level orbits. The electrons can absorb a share of energy to transition from the low energy level to the high energy level orbit, or release a share of energy to return to the low energy level orbit. Energy can be transferred between material particles, and can also radiate a share of energy into vacuum space. The so-called movement of electrons is actually the electric field vibration wave of a certain frequency. When an electron radiates the kinetic energy of its electric field vibration wave into the vacuum space, it is called object luminescence. The atomic structure and electronic orbit determine the energy that can be released or absorbed by electrons. Only light of specific wavelength can be absorbed by electrons. This characteristic makes light have many special functions such as refraction, reflection, transmission, absorption, etc.

The object emits light, that is, the material particles radiate their kinetic energy into the vacuum space by means of electromagnetic vibration. After the vacuum space receives this energy, it becomes the vibration of the electromagnetic characteristics of the space itself. When light propagates in space, it is the vibration of the electric field and the magnetic field of the space body. Since the vacuum space is the "absolute equilibrium state" of the electric and magnetic fields, the vacuum space can carry this energy by the vibration of the electric and magnetic fields. The object emits light, as if a stone is thrown into the calm lake, which can cause ripples on the lake. Electromagnetic wave is the ripple of electromagnetic characteristic fluctuation of "vacuum space".

When energy radiates into space, the electromagnetic field in space carries this energy. After the space carries this energy, the first thing caused is the fluctuation of the space electric field, which becomes the vibration vector of the space electric field. Because vacuum space is an "absolute equilibrium". After the electric field fluctuates, in order to maintain its absolute balance, the vacuum space can only offset the electric field fluctuation by the equivalent change of the magnetic field. This equivalent relation can be given by integral form of Maxwell equations. The equation shows that the changing electric field is equal to the changing magnetic field, and the sum of the two is equal to zero.

$$\oint_L E \cdot dl + \iint \frac{\partial B}{\partial t} \cdot ds = 0 \qquad\qquad 6-1$$

Formula 6-1 shows that the sum of the integral of the changing electric field and the changing magnetic field must be equal to zero. The vibration of the magnetic field is similar to the function of the balance weight, and its function is to ensure the absolute balance of the space structure. It is this balance mechanism of space that leads to the coordinated vibration of magnetic field and electric field. Just like a person walking, in order to maintain balance, he swings his right arm when walking on his left foot and swings his left arm when walking on his right foot. The electromagnetic wave of light reflects the "energy conservation", which is a kind of specification of the "vacuum field" to the "motion of matter". It also reflects the "structure balance" which is a kind of norm of "vacuum field" to "material action". When we understand the normative effect of "vacuum field" on all behaviors, we can understand the physical mechanism of light.

When a person is in the gravitational field, his behavior will be affected by the gravitational field. When people walk, original only needs to walk with feet. However, due to the effect of gravity field, when the left foot steps forward, it needs to swing back with the left arm to maintain balance. When the left arm swings backward, the right arm needs to swing forward to maintain the balance of the upper body. When the right foot steps forward, it also needs similar movements to maintain the balance of the body. For walking with feet, arm movements are only used to maintain balance, which is the effect of gravity field on human walking behavior. Only by walking in this way can one maintain the balance of the body's center of gravity.

Originally, the light energy only needs the vibration vector E of the electric field to express, but in order to ensure the absolute balance of space, the vibration of the electric field E is balanced by the inverse vibration of the magnetic field B. The vibration of magnetic field B is offset by the vibration of electric field E . The repeated balanced vibration of electricity and magnetism causes the electric and magnetic vibration waves to propagate along the normal direction perpendicular to the electric and magnetic vibration plane. The so-called "light particle" is the vibration wave of the "equilibrium" of space electromagnetic characteristics.

Obviously, the speed of light in vacuum is not determined by the energy, but by the electromagnetic properties of space. Light is transmitted by the repeated vibration of electric field and magnetic field in vacuum. The speed of light depends on the response speed between space electric field and magnetic field. The rate of change of the electric field depends on the dielectric constant of the vacuum ε_0, The rate of change of the magnetic field depends on the permeability of the vacuum μ_0, Speed of light C and dielectric constant ε_0, Permeability μ_0 The relationship can be derived from Maxwell's equation:

$$C = \sqrt{\frac{1}{\varepsilon_0\mu_0}} \qquad 6\text{—}2$$

The speed of light in vacuum is the response speed of the electrical and magnetic characteristics of vacuum to the changes of electric and magnetic field. The response speed of electrical and magnetic characteristics is the inherent property of vacuum, which is independent of the size of optical quantum energy itself. The velocity C of light in a vacuum is a constant value.

In fact, a "photon" propagates in space, only transmitting a pure energy and nothing else. We have such a picture in our mind. The endless wheat fields in front of us fluctuate with the wind. The gusts of south wind blowing in the face carried away rows of wheat waves. You can see the wheat waves running to the distance. In fact, the position of each wheat has not moved. They just bend down. The electromagnetic wave signal of light is the vibration of the electric and magnetic fields of the space body. The electric and magnetic fields of the space do not propagate forward, but only vibrate in place; just like the ears of wheat bent under the wind. Only the kinetic energy of the electric field vibration is transmitted to the distance.

Taking space as the carrier of light energy can reasonably explain the Doppler red shift of light caused by the relative movement between the light source and the space carrier, and the cosmological red shift of light caused by space expansion.

Light propagating in vacuum is the vibration wave of space electric and magnetic field. A light quantum is a wave packet of electric and magnetic field vibration. In the vibration wave packet, the field intensity distribution is like a rugby shape, that is, the energy is normally distributed in the wave packet. If the wave packets of light collide, then the energy works. The so-called energy effect is the physical effect of the transfer or rebound of the energy carried by the wave packet of light in the collision. This physical effect will cause the wave packet of light to collapse, showing the characteristics of a particle.

When light presents particle characteristics, it is energy that plays a role. At this time, light has the integral property of energy indivisibility. Indivisibility is a rigid feature of energy, which plays an important role in the wave particle transformation of light. The energy has the minimum unit structure. In the collision process of a photon, either the energy of one photon is fully absorbed, or all the energy is reflected or refracted. There will be no half photon or 1/3 photon. This characteristic of light reflects the nature of energy.

Light is a boson, which is an energy particle. Due to the dependent property of energy, pure energy cannot exist. The so-called static mass of light is zero, because the light energy takes space as a carrier. Through the analysis of photons, the characteristics of neutrinos can be easily understood. Neutrino itself is also an energy particle, belonging to the category of bosons. There are only two kinds of particles in the world, one is material particle, and the other is energy particle, that is, fermion and boson.

To distinguish between fermions and bosons, we should not only consider the mass, but also consider the main role of the particle characteristics. The microscopic particles all have wave particle duality, the material component can only show the wave characteristics, and the energy component can only show the particles. The wave particle duality of microstructure represents that they have both material and energy components.

Space can interact with the motion of electrons and receive the energy released by electrons, which makes vacuum space become the "energy pool" of the universe. The "cosmic microwave background" reflects the energy storage of the "energy pool". The space "energy pool" is a whole, and the overall characteristics of the space are determined by the characteristics of the space "absolute equilibrium". This is similar to the "capital pool" of a bank. The funds deposited in place A can be withdrawn from place B. This overall property of space is shown as "energy correlation", which provides the physical basis for "quantum entanglement".

Electromagnetic space is the vibration of space electromagnetic field, which carries the energy of motion. The energy carried by the vacuum space is not limited to the visible light, but is the electromagnetic vibration wave of all frequencies. The cosmic microwave background is only the energy of electromagnetic field vibration in the microwave band carried by space. From radio waves to γ Ray, these are the vibrations of space electromagnetic field. Only when light is regarded as the vibration of space electromagnetic field can we truly understand that light "fills" all space and "penetrates" the whole cosmic time.

Space is the main research object of modern physics. When you regard space as a physical entity, many problems can be solved and many physical laws can be found here. The most fundamental reason for the emergence of some incongruous and even absurd theories in modern physics is the abandonment of space, the largest subject of existence in the universe.

§ 6·4 Double slit interference of light

With the foreshadowing of the previous section, we can discuss the double slit experiment of light. This experiment has a great impact on the theoretical system of quantum mechanics. It not only reflects the nature of light, but also involves the so-called "quantum superposition state". The double slit experiment of light seems to provide a theoretical basis for theology. The uncertainty of microscopic behavior makes an objective world become ethereal. The double slit interference experiment of light seems to prove that the microscopic field is an unreasonable and mysterious world. Different interpretations of the wave function were also the focus of controversy between the Einstein School and the Copenhagen school.

The double slit interference experiment of light is shown in Figure 6.1:

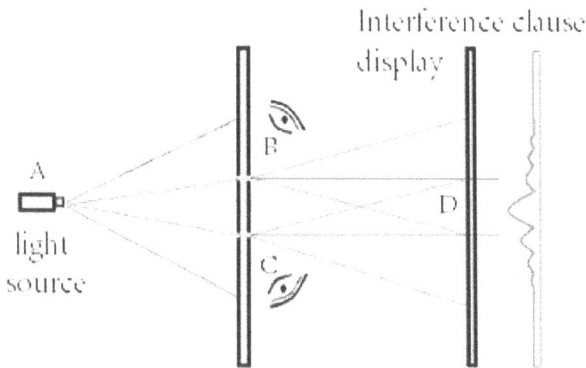

Fig. 6.1

The light source A emits a beam of light, which passes through the double seams B and C and is displayed at the display screen D. The interference between the two projected lights shows that the light is a kind of wave. What happens if you emit only one photon at a time?

Only one photon is emitted at a time, and a light spot is displayed on the display screen D. however, with the increase of the number of emitted photons, the light spot distribution on the display screen still appears fringes similar to wave interference.

In order to find out how photons pass through the double slit, a test probe is installed at the double slit to detect which slit the light passes through. But once the detector head is installed, the light interference fringes disappear immediately. That is to say, when you do not observe, light forms light interference fringes. When you observe, light does not form light interference fringes.

The above questions are: how does a photon interfere with itself? Why does the light interference fringe disappear once the double slit is observed? Since light is a wave, why is only a bright spot displayed on the display screen?

All microscopic particles have wave particle duality, which is the basic feature of microstructure. The light that the light source radiates into the vacuum is electromagnetic wave, whether it is only emitting one light quantum at a time or continuously emitting light pulse wave. This conclusion is proved by experiments and theories in all aspects.

When light travels in space, it is an electromagnetic wave. When it encounters obstacles, light will collide with them. Collision can cause light to refract, reflect or be absorbed by objects. At this time, light presents the characteristics of a particle. The particle characteristics of light are different from those of macro material structure. When the macro material structure collides, it belongs to the direct collision between rigid particles. The refraction and reflection of light, including absorption, is a physical process, which includes the conversion of light from waves to particles. Wave particle duality expresses the characteristics of light conversion from wave to particle.

In the double slit experiment, light is regarded as an electromagnetic wave; if the double slit B and C are not observed and detected, the electromagnetic wave will become two groups of waves after passing through the double slit. The two groups of waves will form light interference stripes at the display screen. The interference of wave can be better explained by Huygens principle.

If the light is observed and detected at the double slit B and C, the observation and detection will destroy the original wave packet structure of the light quantum. When the light passes through the observation point, it will contact with the observation device or the detection signal and collide. The collision will lead to the collapse of the wave packet of light, making the original electromagnetic interaction become a collision of energy properties. This process is also a wave particle wave conversion process. Although the light wave still acts on the display screen in the form of wave after passing through the double slit, the light wave packet will collapse into particles first, and then return to the original wave form in the wave particle conversion process of light. This process destroys the two groups of waves originally formed by double slits. After passing through the observation point, there is only one group of waves reaching the display screen, and one group of waves cannot form light interference fringes.

The so-called observation and measurement actually involve the collision of energy properties. The collision will collapse the wave packets of two groups of light into one energy particle. Because the energy of the light quantum has an inseparable overall rigidity, when the wave packet collides and collapses, the two groups of waves scattered in the space will collapse to the energy action point. When the light energy is not absorbed, the light energy will change from the particle property to the wave form. The process of light energy transforming from particles to waves is similar to that of light emitted by light sources. However, the light wave after the collision has lost the effect of passing through the double slit previously.

The electromagnetic wave packet of a light quantum is actually the space occupied by energy. Where the energy goes is the range of electric and magnetic field vibration. The range of energy action of a light quantum is not a point, a line, or a plane, but a spatial region. When the energy carried by the electromagnetic characteristics of the space body passes through the double slit, it will be divided into two energy action regions. When the energy action area collides, the energy will be concentrated and contracted to the action point. At this time, the vibration of the electric and magnetic fields in the two energy action areas will disappear at the same time. If the energy is not absorbed, the energy will be radiated to the space again in the form of light emitted by the light source. Energy is radiated into space again, that is, particles are transformed into waves.

In the electromagnetic wave packet of light or in the spatial region where energy acts, energy is distributed according to normal distribution. How to understand the normal distribution of energy, let's first analyze the sine wave generated by the self-excited oscillation circuit.

Sine wave is the basic change waveform of electromagnetic field in nature. Any complex electromagnetic oscillation signal can be formed by superposition of sine wave signals with different frequencies and amplitudes. A sine wave can be described by a wave function.

$$E(x,t) = E_0 Sin(K^*x - \omega t) \qquad 6{-}3$$

Where E_0 is the maximum amplitude of the electric field, K wave vector, and K is equal to $2\pi/\lambda$, ω Angular frequency, T time.

Sine wave is the most easily obtained electrical oscillation signal waveform, and it is also a kind of electromagnetic signal waveform that can be naturally generated. When we do not consider the negative half cycle of the signal, the positive half cycle waveform curve of the sine wave is similar to a normal distribution curve, with low at both ends and high in the middle. The normal distribution curve of low at both ends and high in the middle represents the general law of energy change after a new event in nature.

What is a new event? An event in which the original running state has changed. For example, start a car or stop a running car. For another example, self-excited oscillation generates a pulse signal, or an object emits light outward. The world around us is an equilibrium state maintained by the "vacuum field", which is full of the inertia of the original state. If anything wants to change the original state, it must first overcome the original inertia. Therefore, when matter enters a new state of motion, the transition of energy always follows a normal curve. The same is true for electron orbital transitions.

A pulse signal is generated in the self-excited oscillation circuit, and the amplitude has a gradual change from zero to peak, and then gradually attenuates from peak to zero. The outward emission of an object is the energy radiated by electrons, and electrons have a similar process during orbital transition. A photon is a series of electric field vibration waves, the whole wave packet is like a moving rugby, and the energy is also distributed according to the normal curve of low at both ends and high in the middle.

Because a light quantum is the smallest structural unit of energy, there is an "energy correlation" between the electric field vibration wave of light dispersed in a certain spatial region, and the whole wave packet is formed by the rigid action of energy. When the "energy related" wave packet collides, the electric field vibration wave packet dispersed in a certain space region will collapse. It collapses like a deflated balloon, and the energy collapses to a point and becomes a particle.

There are two possibilities for collision: one is that energy is absorbed in collision and transformed into other energy forms, such as heat. The second case is that the light energy is not absorbed after the collision, and the collision causes the electric field amplitude wave of light to shrink into an energy particle. After the collision, the energy particles radiate out in the form of electric field vibration wave, which is similar to the object emitting light. The whole process is a wave particle wave conversion process.

After the collision of light wave packets, that kind of change will occur: it depends not only on the energy structure of light quantum, but also on the frequency of light and the physical properties of the collision object. The energy structure of photon is the distribution of photon energy in space. The energy structure of photons determines the probability of energy transfer when the particle is found or collided.

The energy of a photon is distributed in a certain region, and the integral of the energy in this region space should be 1. Because all energy is equal to the energy of a particle. This particle exists in this region, and the probability of finding a particle in this region is 1. This is that the wave function expressing a "quantum state" must satisfy the normalization condition.

The energy of a particle is distributed in a certain spatial region. At time t, the probability of finding a particle at a certain point in space is related to the energy distribution density of the particle at time t. The density of energy at a point in the space region, which means the probability of energy transfer in the collision process if a collision occurs at that point. The greater the energy density in a small area, the higher the probability of energy transfer after collision.

Similarly, if we use physical methods to detect a small region, the energy density in the region represents the probability of finding particles in the region. The probability of finding particles in this region is directly proportional to the energy distribution density. The greater the energy density distributed in this region, the higher the probability of finding particles. If no particles are detected in a small area, it does not prove that the density of distributed energy at the point is zero, but that there is no energy transfer at the point during detection.

In a certain spatial region, the energy of a photon at time t is distributed according to the normal curve, which means that at this time, the probability of finding particles in a certain spatial region is also distributed according to the normal curve.

In the double slit interference experiment, if the experimental device has only one slit, the stripe on the display screen is a spot with dark on both sides and bright in the middle, which is distributed according to the normal curve. If it is a double slit experiment, and the light interference fringes have been formed, not only the fringe spot of each light interference has a normal curve distribution, but also the overall spot composed of each fringe has a normal curve distribution. In the middle of the whole pattern, we have the greatest probability of finding particles. On the contrary, the dark spot region is the region with the weakest energy distribution, and the probability of finding particles is the lowest.

§ 6·5　Physical significance of Schrodinger equation and wave function

Now we can discuss the Schrodinger equation. The physical meaning of Schrodinger equation is a controversial physical problem in the development of modern physics. Without studying the "balance principle" and the normative role of "vacuum field" in microstructure, we can-not really understand the physical meaning of Schrodinger equation.

Newton's second law of motion reflects the relationship between the acceleration of the movement of macro objects and the gravitational force, and Schrodinger's equation reflects the relationship between the acceleration of the movement of micro structures and the force. What is the motion acceleration of microstructure? What is the force that causes the acceleration of microstructure movement? When we understand these two problems, we can fully understand the physical meaning of Schrodinger equation.

Although the movement of matter is ever-changing, when we understand the relationship between various physical laws, we will find that there are only two forces in nature, one is the universal gravitation caused by mass, and the other is electromagnetic force. The weak force is the universal gravitation generated by the quality of the microstructure, while the strong force is the electromagnetic force in the microstructure.

Microstructure is a "balanced structure" of the community of "material action" and "motion energy". Under the "vacuum field" specification, the mathematical form describing the equilibrium structure of weak force is the mass energy equation. Under the "vacuum field" specification, the mathematical form describing the electric and magnetic balance structure is the Schrodinger equation.

In microstructure, electromagnetic force is a state function, and the magnitude of electromagnetic force will change with the change of microstructure motion state. Since the microstructure is an "equilibrium structure" with the same body of "material action" and "motion energy", the change of electromagnetic force means the change of equivalent energy. Schrodinger equation expresses the mathematical expression of the change of the action state of electromagnetic force and the corresponding energy change.

Microstructure is a non-neutral material organization, and the whole will present electricity or magnetic force. We can use a wave function to describe this tissue. The wave function can be used to describe the motion state in which the magnitude and direction of "material action" change with time. Therefore, the wave function is only related to the mass, electric quantity and magnetic flux of the microstructure, and has nothing to do with the state parameters, energy and momentum of the microstructure. The wave function describes a wave, not the trajectory of a particle. The wave function describes the continuous distribution and change of charge and magnetic flux in space.

Since the effect of mass in microstructure has been determined by the mass energy formula, the wave function does not involve the mass of matter. The mass of the microstructure is m_0, and its corresponding energy is E_0. E_0 is the internal structural energy of the "particle", which does not affect the overall behavior of the "particle" expressed by the wave function. The energy of the overall behavior of the "particle" is the kinetic energy E_K. The "material action" corresponding to E_K is the electromagnetic force of "particles".

The energy corresponding to the electromagnetic force of the "particle" is the momentum P of the overall motion, or the kinetic energy E_K of the overall motion. For a free "particle" not affected by external force, momentum or energy are constant. Therefore, we can use the general form of complex exponential function: $\Psi=\Psi(X)=Ae^{ikx}$ to describe this free particle. When the particle is in a bound state, the momentum P of the particle or the kinetic energy E_K of the overall motion is no longer a constant. For the microstructure whose momentum P or kinetic energy E_K is not constant, it can not be described by general wave function.

The wave function of a "particle" in a "bound state" must satisfy the Schrodinger equation. Schrodinger equation gives the physical relationship between "the action of matter" and "the movement of matter" in the physical system described by wave function when the particle is in a bound state. We take one-dimensional time-free Schrodinger equation as an example to analyze its physical meaning.

Formula 6-5 is one-dimensional time-free Schrodinger equation:

$$[-\frac{\hbar^2}{2m}\partial_x^2+V(x)]\Psi(x)=E\Psi(x) \qquad 6—5$$

For plane waves $\Psi(x)=e^{ikx}$ has:

$$\partial_x^2\Psi(x)=-k^2\Psi(x) \qquad 6—6$$

from $P=h/\lambda$ $p=2\pi\hbar/\lambda$ $P=\hbar k$

$$P^2=-\hbar^2\partial_x^2 \qquad 6—7$$

therefore:

$$[-\frac{\hbar^2}{2m}\partial_x^2+V(x)]\Psi(x)=[-\frac{P^2}{2m}+V(x)]\Psi(x) \qquad 6—8$$

from $P=mV$ $P^2=m^2V^2$

$$\frac{P^2}{2m}=\frac{1}{2}mV^2 \qquad 6—9$$

From formula 6-9, the first term $P^2/2m$ in formula 6-8 is kinetic energy, and the second term $V(x)$ is potential energy. Formula 6-5 one dimensional time-free Schrodinger equation expresses:

$$[-\frac{\hbar^2}{2m}\partial_x^2+V(x)]\Psi(x)=E\Psi(x)$$

Kinetic energy + potential energy = total energy

Schrodinger equation reflects the energy composition of a microphysical system, but it does not only reflect the energy composition. Newton's second law of motion describes the motion law of macro objects, and Schrodinger equation describes the motion law of microstructure. We can put $\Psi(x)$ The wave function is regarded as a physical system. Two forces T and U act on the system. One generates kinetic energy and the other generates potential energy, which is equal to the total energy E in the physical system, as shown in Figure 6.2.

$$[-\frac{\hbar^2}{2m}\underset{Force}{T} + \underset{Force}{U}]\Rightarrow\Psi(x) = \underset{Energy}{E}\Rightarrow\Psi(x)$$

Fig. 6.2

T stands for an external force acting on a physical system. The external force here is to find the second-order partial derivative of the wave function of this microscopic system. In Newton's mechanical system, the first partial derivative dx/dt of displacement x with respect to time is the velocity of a particle v. The second partial derivative d^2x/dt^2 is the acceleration a of the particle motion. In the Schrodinger equation, the force T acting on the physical system is the same as the gravitational force F in Newton's second law F=ma . Both kinds of forces mean that a force causes a physical system to produce accelerated motion. The second-order partial derivative of the wave function is equivalent to a force acting on the physical system described by the wave function, making the physical system produce an acceleration motion. In macro structure, the object of force is a particle. The force in Schrodinger equation acts on the microscopic system. In macro structure: the force causes the particle to produce acceleration motion. In microstructure: as long as the wave function describing the microstructure satisfies the Schrodinger equation, the motion state of the physical system described by the wave function is the motion of acceleration.

In the atomic structure model, the force U is the potential energy of the potential well, representing the gravitational action of the positive charge of the atomic nucleus. If force U is the gravitational action of positive charges, the relationship between force U and force T shows that force T is the physical effect of negative charges of electrons themselves. The so-called physical effect of the negative charge of the electron itself is that because of the negative electric property of the electron, the electron must move in the potential well in an acceleration motion mode. The negative charge of the electron itself also represents a force in the microstructure.

The force U is easy to understand and belongs to the potential energy of positive charges distributed in the potential well, where electrons can convert the potential energy into kinetic energy. What kind of force is force T ? Is it the negative charge of the particle itself ? That is to say, the electron generates a force on itself in the potential well, and this force causes itself to generate acceleration. How to understand this?

In the micro field, in addition to the influence of mass on the structure of matter, particles also need to maintain a balance between two aspects. One is the balance of interaction between individuals.

Such as the balance between positive and negative charges. The other is the balance of individual structure itself. For example, electrons present negative charges. According to the specification of "vacuum field", the existence of any substance must be a "equilibrium structure", or at least an "equivalent equilibrium state". Because the electron is negatively charged, it must use its motion characteristics to counteract its negative electricity characteristics. Moreover, the negative electric force of the electron is a force of the same nature as Newton's gravity F, which needs to be offset by the motion of acceleration nature.

Therefore, the motion in the microstructure is a motion of acceleration "nature", and the force of motion comes from the normative action of "vacuum field". Schrodinger equation describes the mathematical relationship between the gauge action of "vacuum field" and the acceleration motion. That is, for a wave function describing the microstructure, the second partial derivative is obtained. As long as the wave function satisfies the Schrodinger equation, the motion state described by it is the motion state of the acceleration "property" of the microstructure under the "vacuum field" specification.

The reason why the "vacuum field" wants to regulate the movement of a microstructure is that the "particles" show electrical or magnetic characteristics externally, and the "particles" need to offset this effect by accelerating movement. A microstructure is a balance system formed by the electric and magnetic effects of the "particle" itself and the acceleration motion of the "particle" itself. Can this theory be verified? This can be done by imitating Einstein and doing an ideological experiment, which can be verified by comparison.

In "gravitational space", the interaction of macro particles is gravity, and the mass presented is "gravitational mass". If the particle is deep in the universe and there is no mass celestial body around, the role of the particle is the normative role of the "vacuum field". The quality presented is "inertia mass". Macroscopic particles will remain stationary or move in a straight line at a uniform speed in vacuum.

Microscopic "particles" are bound by electric field in the potential well, which is similar to the "gravitational space" of macroscopic particles. The role of micro "particle" in the potential well is potential energy. If the $V(x)$ potential energy term is removed in formula 6-5, the micro "particle" is similar to a macro particle in a vacuum and is not bound by any electric field. What state will the "particle" be?

The "particle" without force in the thought experiment is different from the free "particle", which is the state after absorbing energy. In the thought experiment, the "particle" without force is still in the original motion state. If the $V(x)$ potential energy term is removed from formula 6-5, the conditions for the establishment of the equation will not be affected and the mathematical and logical relationship of the equation will not be destroyed. The content expressed by the equation is the natural state of a micro "particle" in vacuum space. At this time, the "particle" is only regulated by the "vacuum field".

Since the mass m_0 and the corresponding energy E_0 of the micro "particle" are the physical parameters of the "equilibrium structure" within the "particle", it does not affect the overall behavior expressed in the equation. The Schrodinger equation does not involve the mass of microscopic "particles". This also shows that micro "particles" have no "inertial mass" in vacuum space. In the microscopic field, the gauge of "vacuum field" to universal gravitation is only expressed by the mass energy equation. Schrodinger equation only expresses the gauge of "vacuum field" on electromagnetic force.

For plane wave $\Psi(X)=e^{ikx}$ to describe a microscopic "particle", its natural state after being regulated by the "vacuum field" in vacuum space is:

$$[-\frac{\hbar^2}{2m}\partial_x^2]\Psi(x)=E\Psi(x) \qquad\qquad 6{-}10$$

In the square brackets of formula 6-10 is the force T, which expresses not only the normative action of "vacuum field", but also the electrical and magnetic action of "particles". It is also the "material action" of matter. E is energy, representing "the movement of matter". The expression of formula 6-10 is the "equivalence principle", and "the action of matter" is equal to "the movement of matter". That is, $[-\frac{\hbar^2}{2m}\partial_x^2]=E$.

Schrodinger equation clearly expresses the "equivalence principle" in the micro field. The equivalent relation in physics is regulated by "vacuum field". When there is no other action in space, only the gauge action of "vacuum field", Schrodinger equation tells us that "the action of matter" is completely equivalent to "the movement of matter".

When talking about the "equivalent principle" earlier, we did not emphasize the environmental conditions for the establishment of the "equivalent principle". Schrodinger equation gives that "the action of matter" and "the movement of matter" are completely equivalent when there is no other material action around in a vacuum environment. If there is the action of other substances around, the energy of material movement is the sum of the actions of various substances, including the characteristics of "particles themselves", which is the physical meaning of the general form of Schrodinger equation.

The V(x) potential energy in formula 6-5 reflects the action of other substances in space, and formula 6-10 reflects the action of microscopic "particles" themselves, which is also the action of vacuum space. Vacuum space is the basic space of all material existence. Formula 6-10 has universality and is the basic law of material existence. All existence must meet formula 6-10. The physical relationship expressed in formula 6-10 is simple and clear. Whether there is a V(x) potential energy term in the equation does not affect the physical relationship described by the Schrodinger equation.

If wave function $\Psi(x)$ Represents a plane wave physical system. If the physical system wants to exist in vacuum space, it must meet formula 6-10 at least. In formula 6-10, E is energy, representing "the movement of matter". Corresponding to E is ∂_x^2. It represents the normative function of "vacuum field". $-\dfrac{\hbar^2}{2m}$ can be regarded as transformation coefficient.m in the transformation coefficient is the mass of the particle. Now let's analyze the physical meaning of $\hbar$.

The gauge action ∂_x^2 of "vacuum field" also represents the electricity and magnetic force presented by "particles". Electromagnetic force is "the action of matter", which is expressed as a wave. The relationship between "action of matter" and "energy" is expressed by the "energy" carried by the "wave", which is given by Planck constant.

Planck constant is not only a physical parameter, but also an energy unit. The energy carried by a material wave in one cycle is the Planck constant. In the structure of material wave, the energy carried by material wave is fixed, which reflects the quantitative relationship between the two subjects of "material action" and "energy".

The whole microstructure is a wave packet. The energy contained in the wave packet depends on the frequency of the wave. The energy corresponding to a period T of the wave is h, and the total energy of the wave packet is the frequency v Product of and h, $E=hv$. In microstructure, the energy carried by waves cannot be described by parameters such as mass, displacement and velocity. It involves the basic rules of material structure.

The momentum of the macroscopic particle is P, $P=mv$. P_λ Is the momentum of the wave. Momentum of wave $P_\lambda=h/\lambda$. The physical meaning of wave momentum is: the energy of a material wave in a period divided by the wavelength. The momentum of the wave describes the energy of a certain frequency material wave in a unit length of space. It is equivalent to the momentum P of particles, but it has different physical connotation. The momentum P_λ of the wave reflects that the energy of its material wave is inversely proportional to the wavelength. The shorter the wavelength, the higher the energy.

Planck constant h is the energy of a period of a material wave. wavelength λ Is the distance of a cycle. One period is 2π. Divide the energy of the period by 2π, i.e. $h/2\pi=\hbar$. $\hbar$ is the smallest resolution unit of the momentum P_λ of the wavelength. $\hbar$ represents the minimum length unit of the wave and the momentum in the minimum length unit. A length distance x less than $\hbar$, or a momentum P_λ less than $\hbar$, is indistinguishable. This is like a measuring ruler. The smallest scale is 1, and its resolution error is greater than 1/2 scale. If the displacement of particles is used to measure the length distance, the possible measurement error is Δx. The component of momentum P_λ in the x direction is $P_{\lambda x}$. The measurement error is $\Delta P_{\lambda x}$. Then the product of the error of displacement and the error of momentum $\Delta x * \Delta P_{\lambda x} \geqslant \hbar/2$.

Newton's second law of motion reflects the motion law of macro objects, and Schrodinger equation reflects the motion law of microstructure. The force acting on the motion of a macroscopic object is universal gravitation. The force of motion in microstructure is the electric and magnetic force of "particles". Macroscopic objects have "inertial mass" in vacuum. Microscopic "particles" have no "inertial mass" in vacuum. Macro objects are regulated by the "vacuum field" in vacuum and keep stationary or moving in a straight line at a uniform speed. Microscopic "particles" are regulated by the "vacuum field" in vacuum and move with "nature" of acceleration. The reason why a macroscopic object stands still in a vacuum or moves in a straight line at a uniform speed is because of the "conservation of energy". Micro "particles" move in a vacuum with the "nature" of acceleration, which is based on the "balance" of force in the material structure.

Schrodinger equation has clear physical meaning and wave function $\Psi(r,t)$ It only describes two things, electric field or magnetic field. If the described "particle" is an electron, the equation describes the continuous distribution of electric field in space. For example, an electronic cloud. Why did a wave function with clear physical meaning have great controversy in history? Schrodinger said: "The de Broglie wave of the electron describes the continuous distribution of electric quantity in space". Why does this theory encounter difficulties in research and experiment? Why is a "probability amplitude" used to represent the wave function?

In fact, Schrodinger's view is not wrong. The wave function describes the continuous distribution of electric field in space. Moreover, the continuous distribution of electric field in space has the motion characteristics of acceleration. When the physical system has the motion characteristics of accelerated motion, it is equivalent to an energy E acting on the physical system. The continuous distribution of electric field with acceleration motion characteristics in space is completely equivalent to the continuous distribution of energy in space.

The wave function describes the continuous distribution of electric field in space, which can be replaced by the continuous distribution of energy in space. Use $\rho_e(r,t)$ to represent the evolution of energy density in space with time, if the wave function of a three-dimensional Schrodinger equation $\Psi(r,t)$ satisfies Schrodinger equation, then:

$$\rho_e(r,t)=|\Psi(r,t)|^2 \qquad 6\text{—}11$$

And, $|\Psi(r,t)|^2$ when r is integrated from $+\infty$ to $-\infty$, there are:

$$\int |\Psi(r,t)|^2 dV=1 \qquad 6\text{—}12$$

The physical meaning of integral normalization means that the energy distributed in the integral space is the energy of a "particle", and this energy is a conserved quantity.

Why should energy replace the continuous distribution of electric field in space? An electric field wave described by a wave function and changing in space cannot be measured. Measurement is a kind of physical contact, which belongs to the collision of energy property. At the moment of collision, the matter wave will collapse and contract into an energy particle. The process of measurement is also the conversion process of matter from wave to particle.

Although the basic form of the microstructure is a wave, it is a whole and behaves as an energy particle when colliding. When encountering obstacles, there are two kinds of collisions, "elastic collision" and "non elastic collision". There is no energy transfer in the collision, which is called "elastic collision". If a particle with high energy transfers part of its energy to a particle with low energy through collision, the collision is called non elastic collision. Whether it is "elastic collision" or "non elastic collision", the collision is always the effect of energy. At this time, the microstructure fully shows the characteristics of particles.

Microstructure is a kind of wave, because it is distributed in a certain spatial area in the form of wave. Microstructure is also a kind of particle. Besides the rigidity of energy, it also has the quantum properties of matter itself. The so-called rigidity or quantum property means that the microstructure is an inseparable whole. Although matter is distributed in a certain spatial area in the form of waves, when collision occurs, the energy carried by all matter waves will converge to the point where the collision occurs, resulting in "elastic collision" or "non elastic collision".

If we want to observe and measure a microstructure, can we find this material particle in the spatial region of material wave distribution? To observe and measure a microstructure is actually to find or discover this "particle" and obtain its physical information. Whether this "particle" can be observed and measured is an uncertain event. We first rule out the unfavorable factors caused by measurement technology and means, and simply consider the possibility that "particles" themselves are detected.

From the above analysis, we can see that the probability of finding a "particle" somewhere in space is directly related to the energy density of the "particle" distributed there. If the density of energy distributed somewhere in space is given by a wave function, and the wave function is normalized. Then: if all the energy of the "particle" is concentrated in the detected area, the probability of finding the "particle" is 100%. If the detection range is not the space of "particle" energy distribution, and the measurement signal will have no object of action, then we find that the probability of "particle" is zero.

If at time t, near a certain space r, within the dV volume, the probability of finding "particles" is P. At time t, near a certain space r, the energy distributed in the volume of dV is E_ρ. The total energy of "particles" is E. Then:

$$P = \frac{E_\rho}{E} \qquad 6\text{—}13$$

The energy distribution density of a "particle" is $\rho_e(r,t)$ indicates. If at time t, the energy distributed in dV volume near a space r:

$$E_\rho = \int \rho_e(r,t)dV \qquad 6\text{—}14$$

As can be seen from formula 6-11, $\rho_e(r,t)=|\Psi(r,t)|^2$. Therefore, the energy distributed in the dV volume near a certain space r at time t:

$$E_\rho = \int |\Psi(r,t)|^2 dV \qquad 6\text{—}15$$

Since the wave function satisfying the Schrodinger equation is normalized, the total energy E of the "particle" is equal to 1. Therefore, the probability P of finding "particles" in the dV volume near a space r at time t is:

$$p= \int |\Psi(r,t)|^2 dV \qquad 6—16$$

From the above analysis, it can be seen that Max Born probability interpretation of wave function is also correct from the experimental effect alone. The square of the wave function mode, which represents the density of the occurrence probability of "particles" near r at time t. The probability p is equal to the square of the wave function modulus multiplied by this volume element. But we should have a clear concept of the physical relationship. The so-called "particles" originally exist in the form of waves. Finding a "particle" at a certain point in space is because the energy absorbed by the "particle" in the process of measurement or collision makes us know its existence. The probability of discovering "particles" is actually the probability of energy transfer.

§ 6·6 "Quantum entanglement" and "energy correlation"

We now know that the essence of gravity is the repulsive force of space. This is a macro effect of "mass matter". At the micro level, quality cannot be understood as the repulsive force generated by space. Micro particles also have mass, and the mass of micro particles also expresses the interaction between particles and "vacuum field", but this interaction is the interaction at the micro level.

The electromagnetic force will also be different at the macro level and micro level. The electromagnetic force in the microstructure is more magnetic polarity effect. Because the electromagnetic force is the force of the material itself, the electromagnetic force in the structure is regulated by the "vacuum field" in addition to the electromagnetic interaction. On the one hand, the "vacuum field" maintains the balance of its electromagnetic structure, on the other hand, it also keeps the space flat and straight. These two properties of space can be described by "absolute equilibrium". "Absolute equilibrium state" reflects a space state to be maintained in vacuum space. Therefore, the "vacuum field" not only regulates the morphology and structure of the microstructure, but also restricts the behavior of microscopic particles in space.

The space of "absolute equilibrium state" not only regulates the vector balance of forces in the microstructure, for example, one electron in the orbit rotates leftward, and the other must rotate rightward; It also involves the symmetry of spatial geometry. For example, the S track is a central symmetric spherical structure, while the P track can only be a dumbbell shaped structure symmetric to the central axis.

Since vacuum space is a whole connected by "material field", what "vacuum field" regulates is the overall balance of space. For the vector force in space that is symmetrical but separates two places, the "vacuum field" can span the space distance and maintain the symmetry and balance of the two forces. For example, the magnetic field generated by the earth, one pole can be in the South Pole, and the other pole can be in the North Pole. The magnetic field can span the space distance, making the magnetic line of force closed. What are magnetic lines of force? Magnetic force lines are directional forces, that is, vector forces with directivity. Although the Earth's magnetic field belongs to the macro effect, it shows that space can transfer not only the energy of "material movement", but also the vector of "material action" force. This property of "absolute equilibrium state" in space is the basis of entanglement between quantum.

Maxwell's equations reveal some physical characteristics of the "absolute equilibrium" in vacuum space. When there is an electric field vibration vector E in space, there must be another magnetic field vibration vector H that can offset this electric field vibration vector. Similarly, if an electron is excited in the vacuum space, a hole will be left in the vacuum space. This means that when a negative electron is excited, a positron will be produced. Because the vacuum space itself is the "absolute equilibrium" of electromagnetic interaction.

When the energy carried by an electric field vibration vector E in space is absorbed, the magnetic field vibration vector H will also disappear, because the vacuum space has recovered its equilibrium structure. For photons propagating in vacuum space, what space transmits is only pure energy.

It is generally believed that magnetic fields are generated by electric fields and positrons by negative electrons. But the fact is: the magnetic field is not generated by the electric field, and the positron is not generated by the negative electron. Magnetic field or positron is because vacuum space is an "absolute equilibrium system". When a negative electron is lost, a positively charged hole will be generated. When an electric field vibration vector is generated in space, the opposite magnetic field vibration vector will be generated at the same time.

The electric field vibration vector E is related to the magnetic field vibration vector H , and there is also a correlation between negative electrons and positrons. This correlation is determined by the equilibrium characteristics of vacuum space. There are two kinds of correlations in the "vacuum field" specification, one is the correlation of general universality, the other is the correlation of specific objects.

If a certain physical quantity A , it may have two motion states. We use A_0 and A_1 to represent two different motion states of A respectively. Another physical quantity, B , may also have two motion states. We use B_0 and B_1 to represent two different motion states of B . If according to the specification requirements of "vacuum field": if physical quantity A is to be in A_1 motion state, B must be in B_0 state. If is to be in B_1 state, A must be in A_0 state. Under the gauge action of "vacuum field", the mutual correlation between A and B determined by two physical quantities is called entanglement.

The norm of "vacuum field" to "material action" is "structural balance". Any "state" described by wave function is an "equilibrium state", and the "superposition state" formed by "state" and "state" is also an "equilibrium state". If a "superposition equilibrium state" composed of "state 1" and "state 2" cannot be disassembled into two independent "equilibrium states", it is said that "state 1" and "state 2" have an entanglement relationship, and the two sides are "entangled states" with each other.

In quantum mechanics, when more than two particles interact with each other, they form an association. Because the properties of a single particle have been integrated into the overall properties. It is impossible to describe the properties of individual particles separately, but only the properties of the whole system: this phenomenon is called Quantum Entanglement.

Quantum entanglement is a physical phenomenon of the gauge action of "vacuum field". The specification of "vacuum field" requires that every physical existence must be a "balanced structure". If only the combination of A_1 and B_0, A_0 and B_1 can build an equilibrium system, the two physical quantities of A_1 and B_0 or A_0 and B_1 will become associated physical existence. This correlation is different from the correlation between electric field, magnetic field, negative electron and positron, and its correlation object is a unique, specific and mutually directional physical object.

All existence of the visible universe is a balanced structure, which is determined by the balanced characteristics of space. The propagation of light in vacuum actually takes advantage of the balance of the electromagnetic characteristics of the space body itself. The same is true for the physical mechanism of quantum entanglement. Light travels forward by using the balance between the electric field vibration vector and the magnetic field vibration vector. Quantum entanglement is that two particles maintain the entanglement relationship by using the equilibrium property of the space body.

The vacuum space is the background space of all physical behaviors. The balance characteristic of the whole space body extends the "definition domain" of all physical behaviors to the whole universe. Where the balance of space is involved, the butterfly effect will appear. Space will not tolerate the slightest asymmetry and imbalance. The situation awareness capability of space is instantaneous, and the feedback information can be transmitted without any loss. Two entangled particles thousands of miles apart: when the motion state of one particle changes, the motion state of the other particle will change instantaneously. This is a kind of equilibrium maintained by two entangled particles under the gauge of "vacuum field". The force that maintains equilibrium does not come from particles, but from space.

Next, the physical principle of "quantum entanglement" is analyzed through the mechanical structure of optical quantum entanglement. The essence of light is an energy transmitted by electromagnetic waves. Electromagnetic wave has electric field vibration vector and magnetic field vibration vector. The electric field vibration vector represents the energy of light, and the magnetic field vibration vector only plays the role of balance weight. The criterion of "vacuum field" to "material action" is "structural balance". There are two kinds of balances in the electromagnetic wave of light, one is the "equivalent balance" of light energy and electric field vibration vector, and the other is the balance of magnetic field vibration vector and electric field vibration vector.

In vacuum space, the electric field vibration vector representing light energy can be decomposed into two orthogonal components. If the linearly polarized light of electric field vibration vector E is decomposed into vertical o component and horizontal e component by BBO crystal. The linearly polarized light decomposed into two vertical components and the synthesized direction vector must conform to the mechanical characteristics of the original electric field vibration vector. If the polarized light of the vertical o component of the two polarized lights turns into horizontal vibration, the normative action of the "vacuum field" will inevitably turn the e component into vertical vibration. Although two paths of polarized light may be transmitted thousands of miles away, two paths of light must be an orthogonal vibration. Only in this way, the vibration of the electric field in the two light paths will not affect the overall balance of the space.

The energy distribution of light quanta is not a point, but a certain space area. The probability of finding the particle at a certain point in the space area is related to the energy density of the point. It is found that particles are "inelastic collisions". "Inelastic collision" is the process of energy transfer. The so-called probability is actually the possibility of energy transfer of photons. The energy absorbs all the energy of light quanta by the collision of a point, which lies in the correlation of energy existence. The energy of a photon is a whole, either fully absorbed or not. We call the integrity of energy "energy correlation".

A quantum of light is the smallest indivisible unit of energy. If the energy of a photon is divided into two parts and sent to two different places, a and B. At this time, the photoelectric field vibration waves transmitted to a and B belong to the same photon and share the same energy. Although there is no entanglement of vibration direction vector between the two electric field vibration waves, there is still energy correlation.

Light propagating in space is an electromagnetic wave, and a quantum of light is a complete electromagnetic wave package. The electromagnetic wave packet is composed of many sine waves of electric field vibration. Through double slit or half lens, the electromagnetic wave of a photon can be divided into two parts and transmitted to different places. If the electric field vibration wave transmitted to point A accounts for 30%, and the electric field vibration wave transmitted to point B accounts for 70%; then the energy carried by the corresponding electric field: point A accounts for 30% and point B accounts for 70%. What happens if the electric field vibration wave with only 30% energy collides at point A?

For the electric field vibration wave with only 30% energy, will the light energy transfer when the collision occurs? If the collision is a detection behavior, the transfer of light energy means that the particle is found during detection. Otherwise, the particle cannot be detected. For light particles with only 30% of the total energy, it is a probability event to detect or find them. Due to the minimum indivisible integrity of energy, it is impossible to find a particle with only 30% energy here. Either it cannot be detected or a particle with 100% energy is found. What is the physical significance of 30% energy here? 30% energy represents the probability of finding it here.

The size of energy represents the possibility of energy transfer in case of collision. 30% of the energy represents that the probability of finding particles here is 30%. A photon is transmitted to two places, but the energy carried by the electric field vibration wave in both places is the same photon. And the photon is the smallest energy unit that cannot be divided. There is an "energy correlation" between the energy carried by the electric field vibration waves in the two places. The link connecting the energy of the two places is the vacuum space. For the energy of motion, the whole space seems to be the same "energy pool"; we can get all the energy from point A or point B. Point B has 70% energy, and 70% represents the priority of point B to obtain all energy.

Since the energy transmitted to the two places is the same photon, when one side collides, the energy wave packet of the other side will also follow the linkage. If the energy wave packet at point A collapses in the collision, the energy wave packet at point B will also collapse. If one party transfers energy, the transferred energy will include the energy of the other party. The linkage of energy between the two places is the characteristic of "energy correlation".

Energy correlation is similar to quantum entanglement, but also a kind of remote linkage. In addition to the factor that energy has quantum characteristics, the space's specification of motion energy makes energy have the characteristics of linkage in different places. The norm of "vacuum field" on the motion of matter is "conservation of energy", which is the same property as the norm of "vacuum field" on the action of matter. The "conservation of energy" and "structural balance" represent the norms of the "vacuum field" on the two main bodies of the world. "Energy correlation" is related to "energy conservation", and "quantum entanglement" is related to "structural balance". They all have the same physical mechanism. "Energy correlation" also reflects the wave particle duality of microstructure.

The normative behavior of the vacuum field on the world's two existing subjects not only maintains the original balance of space, but also standardizes the two physical behaviors thousands of miles away to the same time, the same place and the same event. This allows us to recognize and understand the "absolute equilibrium" of vacuum, its situational awareness and its characteristics beyond time and space from another perspective. Two particles with "quantum entanglement" relationship and two parts of energy with "energy correlation" exist. Through the connection of "vacuum field", they are still a whole though separated thousands of miles away.

Any existence must be an "equilibrium structure", and entanglement provides the possibility of allowing paired nonequilibrium structures to exist. The existence of this correlation non-equilibrium structure provides a method to store information by using entanglement. Nonequilibrium structure can represent a state. After human operation of its order and different combinations of equilibrium states, it can express some information. If such information can be extracted in two physical systems with entanglement, it will provide a means for communication between the two places.

Whether "quantum entanglement" or "energy correlation", its principle can be applied to quantum communication. The real quantum communication is the information exchange between two totally enclosed physical devices. In this technology, preparation and reading are the key of application technology. Preparation is to manufacture and preserve two entangled physical systems. In two closed physical devices, quantum pairs with entanglement are prepared, and the entanglement between quantum pairs can be preserved for a long time. Reading is to read the state information of the entangled state without destroying the quantum entangled state. Only under this premise, the correlation between the two physical devices is unique and undetectable. The communication of information can pass through any material medium and reach any spatial distance. And the transmission speed is not limited by the speed of light.

The entanglement relationship can be two particles with "energy correlation" or "quantum entanglement". The key to maintaining the entanglement relationship is not to destroy its energy structure or its intrinsic properties. Once the energy or intrinsic properties of entangled particles are touched, the correlation will be released. Entangled state is a local non-equilibrium state. As long as the two sides of entangled relationship can maintain the non-equilibrium state on one side, the entangled state on the other side is not easy to collapse.

Entanglement between quantum is a kind of mutual relationship based on "structural balance". For a "superposition state" of multiple "quantum states", in order to maintain the overall "structural balance", there will also be a similar entanglement relationship between "states" and "states". For example, a multi electron atomic structure. In the multi electron atomic structure, this interaction also exists between particles, and Pauli's "incompatibility" principle reflects this relationship.

An atom is a neutral structure. Each particle in an atom presents different characteristics, including motion characteristics. The particles with different characteristics interact and restrict each other to build a balanced system as a whole. In order to maintain balance, the interaction between particles in an atom can be regarded as an entanglement relationship, but this entanglement relationship is only limited to the interior of the atomic structure.

When the atomic structure is in the equilibrium state of electromagnetic action, the electrons outside the nucleus are arranged according to a certain law. We can use a specific wavelength of laser irradiation to make an orbital transition of an electron. Due to the effect of "structural balance", the electrons with orbital transition will quickly radiate photons and return to the original orbit of the electrons.

There is a fixed logical relationship between how many wavelengths of light are used to irradiate and how electrons will transition. This logical relationship will have different reactions in different atomic structures. For the same atomic structure, but because the atoms are in different unbalanced states, there will be different reactions to the same laser. If we can switch and lock a certain logical state of the atom at the same time, and release it when it needs to be released, it is possible to use this logical relationship for information processing.

Select a material whose atomic structure can meet the requirements for the transition mode of electronic orbit. A group of lasers with different wavelengths are used as input information. Using the above principles to make a physical device with multiple logical relationships as the information processing center. This constitutes a so-called quantum computer.

Quantum computers need to have two basic functions, one is storage, the other is computing. Storage is mainly the preparation of "quantum bits". This is how to create an "unbalanced" quantum state. Since the "quantum state" of the natural state is "equilibrium state", how to use technology to bind a "quantum state" and control the "imbalance". This controllable and unbalanced "quantum state" is a "quantum bit".

Another key and difficult point of quantum computer is how to design: the logical relationship between "state" and "state" determined by the "superposition state" of multiple "quantum states". That is, the logical relationship between the action of laser with different wavelengths and the transition of electron orbit. An atomic structure with different quantum numbers is a "superposition state" of multiple "quantum states". As long as we can bind a certain "unbalanced state" of the atom, we can use lasers of different frequencies to obtain the logical relationship we need. It is also possible to irradiate different "unbalanced states" in atoms with lasers of the same frequency, so that the transition of each electron orbit conforms to a certain logical relationship, and output optical signals according to this logical relationship.

Light is just a kind of energy, and the wavelength represents the size of energy. Light of different wavelengths can represent different information. Light of different wavelengths acts on the "superposition state" of multiple "quantum states", and the change of "superposition state" of "quantum state" represents the logical relationship between information. This logical relationship is quantum computing. Obviously, quantum computing does not need to go through binary conversion, and can directly express a certain mathematical logic relationship.

Both quantum entanglement and quantum computing belong to the physical application of the physical mechanism of "structural balance". The norm of "vacuum field" for "material movement" is "energy conservation", and the norm of "material action" is "structural balance". "Energy conservation" and "structural balance" are the basic laws that matter and motion must follow, and are the result of the specification of "vacuum field". "Vacuum field" regulates all physical behaviors in the universe. The so-called "nonlocality effect", "hidden variable theory" and so on can be answered in the normative role of "vacuum field".

§ 6·7 Delay selective quantum erasure experiment

Delay selective quantum erasure experiment is the combination of quantum erasure experiment and Wheeler delay selective experiment.

Wheeler delay selective experiment is considered to be the most mysterious physical phenomenon. In particular, in 1999, five physicists cooperated to complete the extension of quantum erasure experiment - delayed selective quantum erasure experiment, which is regarded as the experiment of "current behavior can change history" or "human consciousness will affect quantum behavior". We now use the physical principles discussed above to analyze these mysterious physical phenomena.

Let's discuss Wheeler's delay selection experiment first. The experimental principle is shown in Figure 6.6:

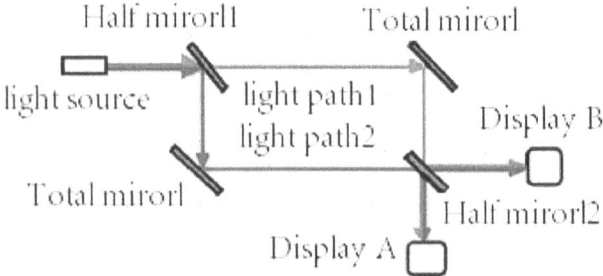

Wheeler delay selection experiment

Fig. 6.6

The light emitted by the light source passes through the half lens 1 and reaches the A detection screen through the full mirror, and the other passes through the half lens 1 and then reaches the B detection screen through the reflection of the full mirror. When the half lens 2 is not inserted, about 50% of the photons reach the A detection screen and about 50% of the photons reach the B detection screen. According to this phenomenon, it is considered that light is a particle, and the probability of reaching A and B is 50% respectively.

When the half lens 2 is inserted, the situation changes. The number of particles reaching A is zero and the number of particles reaching B is 100%. The reason is that each reflection of light will produce a 180° phase difference. After inserting the half lens 2, the light wave reaching A is divided into two paths, one path reaches A after three reflections, and the other path reaches A after one reflection. There is a 180° phase difference between the two paths of light reaching A detector, so that the two paths of light form destructive interference after reaching A detection screen. The two paths of light arriving at the B detection screen are secondary reflection. After arriving at B, the phase is the same, forming phase length interference.

The first case shows that light is a particle, and the probability of light path 1 or light path 2 is 50%. The second case shows that light is a wave, which takes two routes at the same time. Whether light is a particle or a wave, and whether light takes one route or two, needs to be verified by an experiment. This experiment is called delayed selection experiment.

The delay selection experiment is that in the experiment, the half lens 2 is not inserted first, and then the half lens 2 is inserted after the light wave passes through the half lens 1. This is actually carried out according to the experimental scheme 1: when the photon is selected in the form of particle or wave, light path 1 or light path 2, or two routes at the same time, then insert the half lens 2. The experimental results are the same. Light takes two routes at the same time in the form of wave, and interferes with itself.

As we have analyzed in the above article, light is an electromagnetic wave. In both experiment 1 and Experiment 2, light took two routes in the form of waves at the same time. When the half lens 2 is not inserted, the half lens 1 divides the energy of the electromagnetic wave into two parts, accounting for 50% each. Energy distribution also accounts for 50% at A and 50% at B. The energy collides at A and B, and the probability of collapsing into particles is also 50%. After inserting the half lens 2, the electromagnetic wave reaching the A detector forms destructive interference at the half lens 2. The electromagnetic wave arriving at detector B forms a phase expansion interference at half lens 2. Therefore, the energy distribution on detector A is zero, and the probability of finding particles is zero. The energy distributed on the B detector is 100%, and the probability of finding particles on the B detector is also 100%.

Why are we surprised by such a normal physical phenomenon? This is because we did not understand the essence of wave particle duality. Light propagating in space is a wave, and particles are the result of energy after collision.

Next, we will analyze the quantum erasure experiment. The Wheeler delay selection experiment above reflects the essence of wave particle duality. What the quantum erasure experiment reflects is the equilibrium mechanism of "quantum entanglement". The problems reflected by the two experiments are not identical. See Fig. 6.7 for quantum delay erasure experiment:

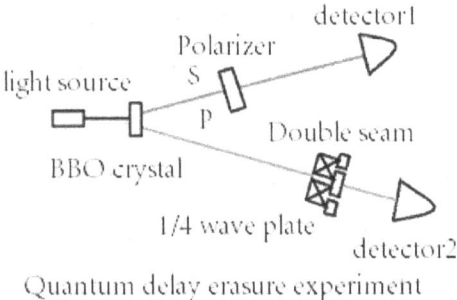

Quantum delay erasure experiment
Fig. 6.7

The quantum delay erasure experiment can be divided into three cases: the first case is that there is no quarter wave plate and polarizer inserted in the diagram, and the optical path P is equivalent to a double slit experiment of light. The interference phenomenon of light can be seen from the detector 2.

In the second case, in order to obtain the path information of light at the double slit, a 1/4 wave plate is inserted before the double slit to obtain whether the light passes through slit 1 or slit 2. When 1/4 wave plate is inserted into the optical path, the interference phenomenon of light disappears.

In the third case, the polarizer of light is inserted into the optical path S. after inserting the polarizer of light, the path information identification of light can be eliminated. At this time, the interference of light is detected at the detector 2.

The original purpose of the experiment was to know how light passes through the double slit, but the result seems to show that the micro world is uncertain. In the microscopic field, the behavior of particles is unpredictable. When you get the path information, the light will not interfere. When the light interferes, you cannot get the path information, and you can only choose one of the two. In fact, what this experiment reflects is the equilibrium mechanism in entangled states.

The light emitted by the light source first passes through the BBO crystal, and one photon is divided into two photons in the entangled state. The entanglement of two photons must meet certain conditions: first, the total "energy conservation" before and after separation. For example, a photon with a wavelength of 351.1 nm can be divided into two photons with a wavelength of 702.2 nm. The other is to maintain the original "structural balance" before and after separation.

If the original electric field vibration vector of the photon is as "a" in Fig. 6.8, it can be divided into S and P components by orthogonal decomposition method. These two component vectors are perpendicular to each other. Before and after separation, the total direction vector remains unchanged. See "b" in Figure 6.8. See Figure 6.8:

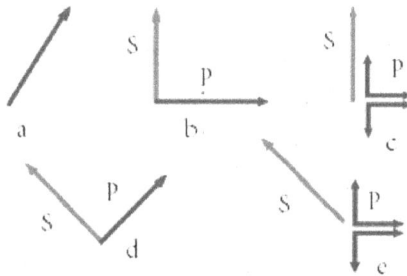

Electric field vector decomposition diagram

Fig. 6.8

The two photons are in entangled state after separation, and the principle they follow is to ensure the original "structure balance". After separation, the vector of the electric field vibration direction of the two beams of light should conform to the vector characteristics of the original force. If one photon vibrates horizontally, the other photon must vibrate vertically. When the electric field vibration vector in the horizontal direction becomes the electric field vibration vector in the vertical direction, the electric field vibration vector in the vertical direction must become the electric field vibration vector in the horizontal direction.

Since the essence of light is an electromagnetic wave, before inserting a quarter wave plate, the optical path of P is a double slit experiment of light. Light can pass through two slits at the same time, thus forming light interference.

After inserting a quarter wave plate into the P-path, the electric field vibration direction vectors of the two quarter wave plates can be drawn in the pattern of "c" in Fig. 6.8. It can be seen from "c" in the figure that the original electric field vibration direction vector P does not match the electric field vibration direction vector of the inserted quarter wave plate. The original electric field vibration direction vector P cannot be decomposed into the electric field vibration direction vector "c" in Fig. 6.8, and the component of the original electric field vibration direction vector P in the vertical direction is zero. After passing through a quarter wave plate, the electromagnetic wave of light has only horizontal amplitude and no vertical amplitude. Therefore, when a quarter wave plate is inserted into the optical path, the light cannot effectively pass through the double slit at the same time, and the interference phenomenon cannot be formed on the detector 2.

After a polarizer is inserted into the S optical path, the polarizer not only changes the polarization direction of the S optical path, but also changes the polarization direction of the P optical path due to the entanglement between S and P. in Fig. 6.8, the electric field vibration direction vector P of "d" has both vertical and horizontal components. The electric field vibration direction vector of "d" can be decomposed into two direction vectors of "e".

After the polarizer is inserted into the S optical path, not only the electric field vibration direction vector of P matches the electric field vibration direction vector of the two quarter wave plates, but also the polarization direction of the S optical path, "d" and "e" are consistent. In the P optical path, the electric field vibration direction vector of the light completely conforms to the electric field vibration direction vector of the quarter wave plate, and the electromagnetic wave of the light can pass through the double slit well at the same time, so the interference of light can be formed on the detector 2.

In the entangled state, the included angle of the vibration direction vector of the two photon electric fields remains unchanged, which conforms to the mechanical structure of the spatial equilibrium state. It is not only the specification requirement of "vacuum field", but also the basis for two photons to maintain entangled state. In the quantum erasure experiment in 1999, the entanglement relationship is no longer simply expressed as "quantum entanglement" between two photons. There is also an "energy correlation" between two photons. Between two photons, there are both "quantum entanglement" and "energy correlation".

The improved quantum erasure experiment in 1999 is shown in Figure 6.9:

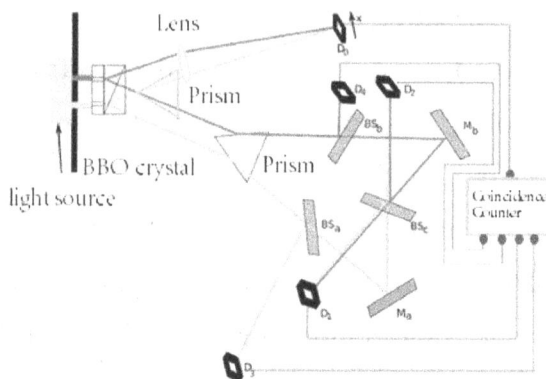

Improved quantum erasure experiment
Fig. 6. 9

In order to facilitate the description, the experimental device is simplified, as shown in Figure 6.10:

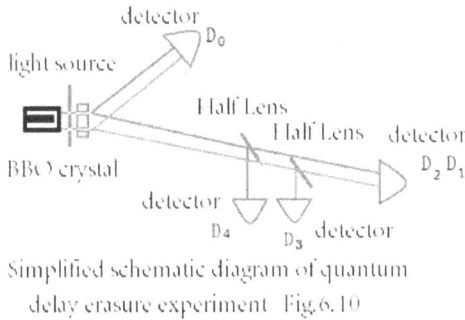

Simplified schematic diagram of quantum
delay erasure experiment Fig.6.10

In the figure, the detectors D_1 and D_2 can be replaced by one detector, because the optical signals they receive come from two gaps at the same time. The physical principles of one detector and two detectors are exactly the same. The simplified schematic diagram can clearly explain that the light signal will not form interference fringes on D_3 and D_4 detectors, and the light can form interference fringes on D_1 and D_2 detectors.

The experimental principle of quantum delay erasure is shown in the figure above. The light wave emitted by the light source is divided into two paths through the double slit, and the two light waves become two pairs of entangled "photon" pairs through the BBO crystal. The "photon" pair in entanglement is divided into two paths, one is "signal photon" and the other is "marker photon". "Signal photon" is sent to detector D_0 to observe its interference phenomenon. "Labeled photons" are sent to detectors D_1—D_4 to observe which crack the "photons" come from.

Experiments show that if D_3 and D_4 detectors that can mark path information receive signals, there will be no light interference fringes on detector D_0. Only when the D_1 and D_2 detectors representing the erase path information receive the optical signal, the detector D_0 will appear the fringe of optical interference. The doubt here is that the distance between the light source and - is greater than that between the light source and . The optical signal has already reached detector before reaching D_1-D_4. Will the later behavior change what has happened?

In order to illustrate this phenomenon, first analyze the working principle of optical signal in the experimental device. Whether a light source emits a "photon" or a beam of light, they are electromagnetic waves. The light wave will reach the BBO crystal through two slits at the same time.

When light passes through the double slit, the light is divided into two parts. But there is an "energy correlation" between the two lights. BBO crystal turns the light passing through each slit, into two entangled "photon" pairs. At this time, there is both "energy correlation" and "quantum entanglement" between each "photon". The energy of each photon is only 1/4 of the original, that is, the energy of the original "photon" is divided into four parts.

The function of BBO crystal is to divide a "photon" into two, and the energy of each "photon" is 1/2 of the original. Because photons are the smallest unit of energy. If a "photon" is divided into four parts, there is not only the entanglement of "material action" but also the correlation of "motion energy" between photons. The correlation between quantum entanglement and energy makes the four parts of energy only have 1/4 of the "photons". No matter where you send them or not, they are a whole.

The energy of "labeled photons" sent to D_3 and D_4 detectors is only 1/4 of the original. D_3 and D_4 are independent detection systems that cannot form light interference between two photons. It is for this reason that there is not only "quantum entanglement" but also "energy correlation" between the "marked photons" received by D_3 and D_4 detectors and the "signal photons" emitted to D_0 detectors.

Both D_1 and D_2 detectors can receive the optical signals of two slits at the same time. Interference can be formed between the received "marked photons". At the same time, there is not only "quantum entanglement" but also "energy correlation" between the "labeled photon" received by D_1 and D_2detectors and the "signal photon" of D_0 detector.

Summarize the above points: there is not only "quantum entanglement" but also "energy correlation" between D_3, D_4 detector and D_0 detector. On D_3 and D_4 detectors, the light cannot form the physical effect of interference fringes and will be reflected on D_0 detector. There is not only "quantum entanglement" but also "energy correlation" between D_1, D_2 detectors and D_0 detectors. On D_1 and D_2 detectors, the light will form the physical effect of interference fringes and will be reflected on D_0 detector. In order to clearly show the relationship between them, according to the above principles, we can redraw the schematic diagram of quantum delay erasure experiment. See Figure 6.11:

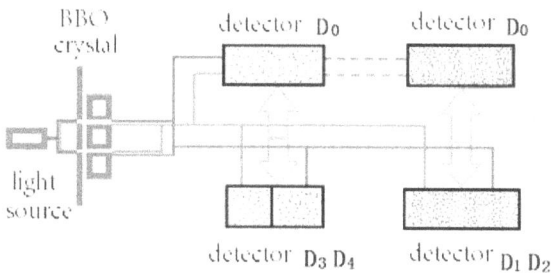

Schematic diagram of quantum erasure
experiment principle Fig. 6.11

It can be seen from the figure that on D_3 and D_4 detectors, the light cannot form interference fringes, which will be reflected on D_0 detector. Similarly, on D_1 and D_2 detectors, the light will form interference fringes and will also be reflected on D_0 detector. For each "photon" without interference received on D_3 and D_4 detectors, the same "photon" will be received on D_0 detector. The same is true between D_1, D_2 detectors and D_0 detectors. The above situation is what we saw in the experiment: when the photons received by D_0 detector have no interference fringes they belong to the reflection of and detectors. When the photon received by detector has interference fringes, it belongs to the reflection of D_1 and D_2 detectors.

The so-called "quantum entanglement" or "energy correlation" means that a quantum behavior will inevitably affect the behavior of the associated quantum. Although the "signal photon" has previously reached the D_0 detector, this signal photon has only 25% of the energy of the original "photon". Before the photons associated with the "photon" energy cannot generate chain response, although collision may occur, this collision belongs to "elastic collision", and energy transfer will not occur during collision. Energy transfer requires "energy related" photon synchronous reaction. "Signal photons" must wait until the chain response of the entangled and energy related photons, that is, after all the energy of the original photons is redistributed, before truly completing the non-elastic collision. The signal detected by the detector is the information of energy transfer.

The focus here is on the difference between non elastic collision and elastic collision. The collision without energy transfer in the collision is "elastic collision", and "elastic collision" will not be displayed on the D_0 detector, but it plays a delay role. The speed of light propagation in the medium is less than that in vacuum, because light has "elastic collision" in the medium. "Elastic collision" slows down the speed of light propagation.

How to prove the above theory? We only need to install a clock timing device in the quantum delay erasure experimental device to record the time of receiving "photons" on each detector. If the received "signal photon" does not interfere, it should be at the same time as the receiving time of "mark photon" on D_3 and D_3 detectors. If the "signal photon" received on the D_0 detector has interference, it should be at the same time as the "mark photon" received on the D_1 and D_2 detectors.

Simultaneity is caused by the "quantum entanglement" and "energy correlation" of two photons. For particles with "quantum entanglement" and "energy correlation", no matter who acts first, the entangled particles must complete energy transfer at the same time. The essence of energy transfer is the collapse of the original optical quantum energy wave packet.

The wave packet collapse can be caused by "photons" on the D_0 detector, or "photons" on the D_1, D_2 detectors or D_3, D_4 detectors. If the photons on the detector collide first, the energy transfer can be completed only if the photons on the D_1, D_2 or D_3, D_4 detectors respond in time. Since response also requires time, there may be differences between active and passive response times.

Another case is that the "photon" on the D_0 detector directly leads to the collapse of the wave packet. Because the wave contains 25% of the energy of the "photon" of the original light source. If 25% of the energy can play a role in the collision, it has a 25% probability of directly completing the energy transfer. However, this energy transfer does not belong to the category of "quantum entanglement" or "energy correlation". It is characterized by that a "photon" can be detected on d0 detector, but the entangled "photon" will not be detected on D_1, D_2 detector or D_3 and D_4 detector.

PART SEVEN

Dark matter

In this chapter, we will discuss "dark matter". The three forms of matter discussed above belong to the theoretical scope of the visible universe. The "physical model" involved in the theory is derived from the hypothesis of the "big bang in space", which is consistent with various known physical laws.

"Dark matter" involves the theoretical model of "big bang in space", but the existence or non—existence of "dark matter" cannot be deduced from the previous theoretical system. Whether "dark matter" exists or does not exist, it will not destroy the integrity of the original theoretical system or affect the self—consistency of the theory. The reason is that "dark matter" does not belong to any of the three material forms.

The "dark matter" discussed here is: if there is "dark matter", according to the theory of the "big bang in space", what kind of "physical *model" will "dark matter" be.*

We should first establish a "physical model" of "dark matter", so that we can have a definite aim. The current means of detecting "dark matter" can be described as "from heaven to earth". There are satellites in the sky and an underground laboratory 2400 meters deep. Although a lot of money has been spent, it may not be fruitful. The main reason is that we do not know about "dark matter". The existing theory holds that the matter in the universe is mainly "dark matter". If the matter in the universe is mainly "dark matter", what impact will this have on the Big Bang theory?

§7·1   Theoretical discussion on " dark matter "

The hypothesis of "dark matter" comes from the observation of the rotation curve of galaxies. According to the rotation curve of each galaxy, the "dynamic mass" of galaxies should be much greater than the estimated and observed mass of celestial bodies. In addition to the rotation curve of galaxies, another evidence of the existence of "dark matter" is the mass light ratio of galaxies. The "rotational mass" of galaxies is much greater than the "luminosity mass" of galaxies, which means that in addition to the objects we can see, there are also "dark matter" that we can't see.

The so-called "dark matter" is that the material does not emit light. We now know a lot about light. Light is an electromagnetic wave in a vacuum. The essence of luminescence is that the object radiates an energy outward, and this energy radiates outward in the form of electromagnetic interaction. "Dark matter" does not emit light involves two questions: "does dark matter have energy? Can "dark matter" radiate energy in the form of electromagnetic interaction?

To answer this question completely, we have to go back to the original theoretical system. The key point here is to fully understand that there are only two kinds of forces in nature, namely gravity and electromagnetic force. The weak force is the universal gravitation in the micro scale range, while the strong force is the electricity and magnetic force in the microstructure. Gravitation is the interaction between matter and space, and only electromagnetic force is the inherent force of matter itself. Gravitational force can make matter gather together, and electromagnetic force can make matter form structure and become electrons, neutrons, protons, nucleons, atoms, celestial bodies, etc. The theory that there are only two forces in nature not only enables us to know the structural laws of ordinary matter, but also deepens our understanding of the spatial structure, which is the key to solving the mystery of "dark matter".

According to the analysis of known facts, of the two natural forces, "dark matter" only participates in the action of universal gravitation and has "gravitational mass". "Dark matter" has "gravitational mass", which shows that it has the general properties of matter and has the same root as ordinary matter. "Dark matter" does not emit light, and its essence is that it does not participate in electromagnetic interaction. Do not participate in electromagnetic interaction because "dark matter" does not have electromagnetic properties. This means that "dark matter" is different from ordinary matter. Ordinary matter not only has "gravitational mass" but also has electrical or magnetic properties, while "dark matter" has only "gravitational mass" and has no electrical or magnetic properties.

In the visible universe, any particle is charged or magnetic, and it must participate in electromagnetic interaction; not participating in electromagnetic interaction shows that "dark matter" itself does not have electromagnetic characteristics. Without electromagnetic properties, it can-not be any of the particles we know. This means that "dark matter" does not belong to the category of matter in the visible universe.

Electricity and magnetic force are the force of matter itself. In addition to forming the structure of matter into electrons, protons and neutrons, electricity and magnetic force can also exchange energy with space through electromagnetic interaction. Light is the energy transferred from material structure to space through electromagnetic interaction. If "dark matter" itself has no electrical or magnetic characteristics, "dark matter" cannot radiate energy into space (does not emit light), nor can it form any material structure. Then we can draw a conclusion that "dark matter" is just a kind of primitive "field" of matter.

Because "dark matter" has "gravitational mass", if "dark matter" is a particle, then "dark matter" has "inertial mass". The "inertial mass" of particles will have kinetic energy and generate corresponding physical effects at the same time. Similarly, because "dark matter" has a "gravitational mass", the "gravitational mass" of particles will cause mutual attraction, resulting in the macroscopic effect of universal gravitation. The universal gravitation of macro effect will form a larger structure through gravitational action in the structure of galaxies with "dark matter" as the main component. At this time, "dark matter" is not only a physical quantity of "gravitational mass". Therefore, "dark matter" cannot be a "particle" in any form.

The form of "dark matter" can only be a "field", because it does not have an "inertial mass". "Inertial mass" describes the conservation of momentum of particle motion. If "dark matter" has "inertial mass", its motion track can be tracked. We can't find any trace of the movement of "dark matter", indicating that it has no "inertial mass".

The mass of "dark matter" in a galaxy is several to dozens of times that of ordinary matter. If "dark matter" has an "inertial mass", it will participate in the cyclic evolution of ordinary matter. Will completely change the trajectory of the galaxy. If the "dark matter" in the Milky Way has an "inertial mass", the solar system will not run smoothly for

4-5 billion years. Since "dark matter" has no physical structure, it is just a "field" in space. It does not have the momentum of macro objects, nor can it transfer the momentum of macro objects. In addition to the gravitational effect, there is no physical quantity that affects the motion of other celestial bodies.

What does "dark matter" mean without "inertial mass"? "Dark matter" has no "inertia mass", which indicates that the macro overall motion of "dark matter" is not regulated by the "vacuum field" and does not follow the "conservation of energy". There will be no moment of inertia or angular momentum. The "dark matter" in each galaxy is just a fog dragged by the gravity of the galaxy. At the macro level, "dark matter" is not regulated by the "vacuum field", and at the micro level, "dark matter" is regulated by the "vacuum field". Participating in the "weak interaction" of universal gravitation in the microscopic field, there is a vibrating motion similar to that of a string.

At the micro level, "dark matter" has a "gravitational mass" and contains energy equivalent to the "gravitational mass". Otherwise, "dark matter" will attract each other and concentrate in the central region, not evenly distributed in the entire galaxy space. The macro level of "dark matter" has no kinetic energy, is not subject to the norm of "vacuum field", and does not abide by the "conservation of energy". At the micro level, "dark matter" participates in "weak interaction", contains the basic energy expressed by the mass energy equation, and is subject to the norm of "vacuum field". This peculiar characteristic of "dark matter" is that it has no physical structure, but is just a kind of "field" scattered in galactic space.

We can explain it by the motion state of an electron. The electron has a "gravitational mass", so corresponding to the "gravitational mass" of the electron, the electron has the energy expressed by the mass energy equation. The electron also has a negative charge, corresponding to the negative charge of the electron. In addition to the energy expressed in the mass energy equation, the electron also has kinetic energy. Kinetic energy is expressed in terms of "inertial mass". "Dark matter" has only "gravitational mass" and no electrical or magnetic characteristics, so "dark matter" has only the energy corresponding to "gravitational mass" and no kinetic energy corresponding to "inertial mass".

"Dark matter" exists in the "vacuum field" and is regulated by the "vacuum field". It has "energy" corresponding to "gravitational mass". If "dark matter" does not have this "energy", dark matter will attract each other and become a mass of matter. "Energy" is the only power that can make matter repel each other and make matter occupy space. The gravitational effect and energy repulsion of "dark matter" are only effective at the micro scale, similar to "weak interaction". Because "dark matter" has no material structure, its "weak interaction" is also different from ordinary matter. The "weak interaction" of "dark matter" only represents the "equivalence relationship" and follows the "mass energy equation" $E=MC^2$, Without quantum properties. Ordinary matter forms matter wave packets under weak interaction, while "dark matter" is similar to a vibrating "string" under weak interaction. This property of "dark matter" is because it does not have electromagnetic properties, and the "material field" of "dark matter" cannot form an organizational structure.

In astronomical observation, we have not found any trace of "dark matter", but only traces of the existence of "dark matter" in the rotation curve of the galaxy. This phenomenon is consistent with the above properties of "dark matter". Why does "dark matter" have the above properties? It starts with the nature of the Big Bang in space.

§ 7·2 "Physical model" of " Dark matter "

"Dark matter" involves the big bang theory. If "dark matter" exists, then the big bang theory must be able to accommodate the "dark matter" theory. Without considering "dark matter", the whole universe is composed of three parts: the visual universe of "equivalent equilibrium state", "non-equilibrium black hole" and "absolute equilibrium" cosmic space. If there is "dark matter", what is the relationship between "dark matter" and the big bang?

We can see that the sun, moon and stars are the visible universe, which is composed of the remaining polarity "material field" and energy from the big bang. Black hole is the form of matter after the "matter field" with polarity loses energy. Space is the form of matter formed after neutralization and annihilation of two material fields with different polarities. The whole visible universe is related to the polarity of the "material field". All the previous discussions are aimed at the visual universe and do not involve the "material field" without polarity. The "matter field" without polarity is not within the scope of matter discussed above.

The characteristic of the visible cosmic matter is the polarity of the "matter field". The three forms of matter are all formed by the polarity of the "material field". The Big Bang in space is the neutralization and annihilation reflection of the polarity of the "matter field". The polarity of the "material field" is shown in the material structure as the positive and negative electrodes of the charge, and the N and S poles of the magnetic field. Between polarities, the same sex repels and the opposite sex attracts. All substances in the visible universe have electrical and magnetic properties. They have magnetism without charge. The material structure in the visible universe is formed by the polarity connection of electromagnetic.

Before the big bang, was there a "matter field" without polarity? If there is a substance in space that has no polarity, or the so-called polarity is closed in the substance itself, making the substance lose its polarity, then such substance cannot participate in the big bang. What kind of "physical model" will a "material field" without polarity be? Does this non polar "matter field" have similar physical characteristics to "dark matter"?

The main characteristics of "dark matter" are: it has mass, participates in the action of universal gravitation, and does not participate in electromagnetic interaction. We now know that matter has three forms: stars, black holes and space. In these three different forms of matter, there are only two kinds of forces, namely universal gravitation and electromagnetic force. Weak force is the expression of universal gravitation in microstructure under the "vacuum field" specification. Strong force is the electromagnetic force in the microstructure under the "vacuum field" specification.

How does gravity come into being? As mentioned earlier, the gravity of matter is actually the repulsive force of space. The repulsive force of space is caused by the influence of mass matter on the flatness and straightness of space. Flat and straight space is an elastic space, and the shape of space can be changed. Due to the existence of mass matter, space has been deformed. After space deformation, a reverse repulsion force is generated. The universal gravitation is the repulsion force of space against mass matter.

Electromagnetic force is different from universal gravitation. Electromagnetic force is a force of "material field". Electromagnetic force is actually the polarity of "material field". Polarity makes an electronic structure negatively charged and a proton structure positively charged. This structure of matter is called positive matter. The polarity of matter field can also make the structure of matter opposite to that of positive matter. The structure of matter opposite to positive matter is called antimatter. If there is another kind of matter in the universe, its "material field" has no polarity, or the polarity has closed itself in the original state and does not show polarity externally. This "material field" without polarity does not have electromagnetic characteristics.

The polarity of the "material field" without polarity is obviously different from that of the "absolute equilibrium" of space. Space is a neutral structure formed by two different polarity "material fields", which is the equilibrium state of electricity and magnetism. The "material field" without polarity itself has no polarity, that is, it is neither charged nor magnetic. It is a "material field" that does not participate in electromagnetic interaction.

What is the difference between nonpolar "matter field" and

ordinary matter with polarity?

1.Nonpolar "matter fields" do not participate in the neutralization reaction of the big bang. The big bang in space is a neutralization reaction of two "material fields" with different polarities. If the "material field" does not have polarity, it will not participate in this neutralization reaction. In the big bang, except for two "material fields" with different polarities, most of the "material fields" themselves do not have polarities. They did not participate in the neutralization annihilation reaction in the big bang. The high temperature environment produced by the big bang makes these nonpolar "material fields" obtain some energy. When space expands, these "material fields" that obtain energy are still scattered in space in the form of "fields" with the expansion of space. These "material fields" are "dark matter".

2.The "material field" of "dark matter" does not have polarity, so it does not have electromagnetic characteristics. Without electromagnetic characteristics, they will not participate in electromagnetic interaction. The result is that in addition to not participating in strong interaction, it cannot form material structure. Whether macro or micro, it is electromagnetic force that makes matter form structure. The difference between positive matter and antimatter is also determined by the electromagnetic properties of matter. Some electromagnetic characteristics determine that matter is a "positive matter" structure, and some electromagnetic characteristics determine that matter is an "antimatter" structure. Some substances do not have electromagnetic properties. Substances without electromagnetic properties cannot form structures. When the "material field" does not have electromagnetic characteristics, there will be no connection between the "material fields". Therefore, "dark matter" can only exist in the form of original "material field".

3."Dark matter" can only exist in the form of original "matter field". On the one hand, it cannot form such material structures as electrons, neutrons and protons, and on the other hand, it cannot participate in electromagnetic interaction. Unable to participate in electromagnetic interaction, which makes it impossible to exchange energy between "dark matter" and space in the form of electromagnetic interaction. On the one hand, "dark matter" cannot absorb the electromagnetic wave energy of light, and on the other hand, it cannot radiate energy through electromagnetic interaction. Because of this, we cannot see them in the spectral range of electromagnetic wave light, so it is called "dark matter".

4."Dark matter" can only exist in the form of the original "matter field", and it cannot be a particle in any sense. "Dark matter" is not a particle, which means that it has no "inertial mass", and also means that "dark matter" has no kinetic energy and will not transfer kinetic energy. Kinetic energy in microstructure is essentially energy corresponding to electric and magnetic characteristics. The fact that "dark matter" has no kinetic energy is consistent with the fact that "dark matter" does not have electromagnetic properties.

5."Dark matter" does not have electromagnetic force, nor does it have energy corresponding to electromagnetic force, which is consistent with the fact that "dark matter" has no material structure.

6.What does it mean that "dark matter" has no material structure and "dark matter" has no "inertial mass"? This means that the "dark matter" in space has no moment of inertia. That is, "dark matter" has no angular velocity, and it will not follow the rotation of the galaxy. But "dark matter" has "gravitational mass", and it will be close to the center of the galaxy. Under the gravitational action of the galaxy, it follows the galaxy to make the revolution of the galaxy. The orbital motion of "dark matter" has no momentum property and is not regulated by "vacuum field".

7.Because "dark matter" has "gravitational mass", it is just a "field". In this way, the "gravitational mass" of "dark matter" causes effects on two extreme levels. One is the interaction at the "field" level, and the other is the overall impact on space. At the level of "field", the "matter field" of "dark matter" interacts with the "vacuum field". This kind of interaction is similar to the weak interaction, which makes "dark matter" have the energy expressed by the mass energy equation. It is this energy that makes "dark matter" disperse in the huge space of a galaxy's gravitational range. "Dark matter" gathers with each other by "gravitational mass", and the energy of the mass energy equation is loosely separated from each other. On the other hand, "dark matter" will affect the flatness and straightness of space as a whole. This will virtually increase the overall mass of a galaxy. This is also the reason why "dark matter" gathers together.

8.Since "dark matter" has no polarity, it has not participated in the evolution of matter from the beginning. Because ordinary matter contains more energy, there will be very intense material movement. "Dark matter" itself does not have energy, and there is no excessive incentive behavior. In addition to not participating in the circulation and evolution of ordinary matter, the existence of "dark matter" in the structure of galaxies can enhance the gravitational effect of galaxies and maintain the stable operation of galaxies.

9.In the evolution cycle of ordinary matter, "dark matter" can neither absorb energy nor release energy. Therefore, "dark matter" is a form of matter that can exist for a long time.

In the early universe, the whole space was full of "dark matter". With the expansion of space and the evolution of galaxies, the "dark matter" gradually formed a cluster organization centered on galaxies under the gravitational action of galaxies. Because the distribution of "dark matter" is affected by the mass and evolution of early galaxies, the "dark matter" carried by different galaxies is also different. In the structure of galaxies, "dark matter" fills the entire interstellar space, and its existence makes the universe stable. The existence of "dark matter" is also conducive to the use of electromagnetic force to form large-scale structure of ordinary matter.

§7·3 Detection of " dark matter "

Since "dark matter" is a "field", how do we detect "dark matter"? The existence of any substance has two main characteristics: one is "the action of matter", and the other is the energy contained in matter. Because "dark matter" has no structure and only participates in weak interaction, "dark matter" only has macroscopic gravitational effect on the whole in the region where it exists. For energy, "dark matter" does not have "inertial mass", momentum, etc. "Dark matter" is a kind of "field", which does not carry high energy. We can't catch their shadows from the high-energy area, nor can we find their traces from the high-energy particle accelerator. Because "dark matter" does not have the physical properties and parameters that ordinary "particles" should have, we cannot detect the existence of "dark matter" by detecting ordinary particles.

"Dark matter" is a kind of "field", and the whole galaxy is surrounded by "dark matter". The vacuum environment of the Milky Way galaxy is full of "dark matter". There is also "dark matter" in the vacuum around us. If we consider the energy contained in "dark matter", the "vacuum energy" originally considered by us may be the energy form of "dark matter" under "weak interaction".

If there is energy in vacuum, vacuum energy may be the energy contained in "dark matter". Vacuum space is the "absolute equilibrium state" of matter, and it will not have any "action of matter" or any "movement of matter". There is no energy in any form in vacuum. If any energy can be detected in vacuum, this energy is the energy carried by "dark matter".

Because "dark matter" has a "gravitational mass", it is subject to the norms of "vacuum field". The regulation of "vacuum field" on "dark matter" is similar to weak interaction. The weak interaction makes "dark matter" contain the most basic energy, which makes "dark matter" like vibrating "strings". These vibrating "strings" float in space like ghosts. Because there are vibrations similar to "gravitational waves" in space, these vibration "strings" will fluctuate with the vibration of the space body. This will cause the difference in the distribution density of "dark matter" energy in space, making it possible for us to detect this difference.

If vacuum energy is detected in vacuum, it should be the energy carried by "dark matter". Another way to detect "dark matter" is to detect the gravitational effect of "dark matter". "Dark matter" has a "gravitational mass" and has a macroscopic gravitational effect on the whole. But they are scattered in space in the form of field, and we cannot detect their existence at a point, or even in a space region. But we can judge the gravitational effect of "dark matter" from two aspects: microscopic detection and macroscopic analysis.

Since "dark matter" is not a rigid body, has no rotational inertia or angular momentum, it is just a fog dragged by the galaxy in the galaxy. "Dark matter" takes part in the revolution of the whole galaxy and forms relative motion with the movement of celestial bodies in the galaxy. "Dark matter" acts as the background space of motion in the space of galaxies. The rotation or revolution of the earth forms relative motion with the background space of "dark matter". The earth's rotation or revolution will also have a dragging gravitational effect with "dark matter", which will produce gravitational differences in different directions of the earth's rotation or revolution. It is possible to detect the change of this gravitational effect through special devices.

The gravitational action of the sun and moon causes tides and slows down the rotation of the earth. If "dark matter" exists, the reason why the earth's rotation speed slows down should also include "dark matter". The sun, moon and other planets can calculate their gravitational action according to the operation law of celestial bodies. Of course, there are also ways to detect the gravitational action of "dark matter".

The above is microscopic experimental detection, and the other is macroscopic theoretical analysis. Macro theoretical analysis is to evaluate the role of "dark matter" as a whole. If there is "dark matter" around the galaxy, it will change the mass of the whole galaxy. According to the rotation curve of the galaxy, the "dynamic mass" of the galaxy can be calculated. Then the "gravitational constant" of the galaxy is modified according to the "photometric mass".

"Gravitational constant" may not be a constant, it may represent some kind of space gravitational environment. The amount of "dark matter" in galaxies may change this parameter. "Gravitational constant" is a constant in Newton's formula of universal gravitation. It can be seen from the formula that the size of gravity mainly considers the mass of two objects, without considering the environmental changes in space. The gravity between two objects must be related to the gravitational environment of the space where the two objects are located. The universal gravitation formula represents the gravitational environment of all space with a constant. When the gravitational environment of space changes, this constant should also change. As for the amount of "dark matter" in different galaxies, the gravitational constant should also change accordingly. The amount of "dark matter" dragged by a galaxy determines the size of the gravitational constant. Different galaxies have different "gravitational constants".

The rotation curve of a galaxy reflects its "dynamic mass". If the "dynamic mass" is obtained according to the rotation curve, then the "gravitational constant" is modified according to the "dynamic mass". If the "gravitational constant" is modified, the "photometric mass" of the galaxy can be consistent with the "dynamic mass" of the galaxy.

It shows that the gravitational action of "dark matter" will change the "gravitational constant" of the galaxy. In astronomical observation, for different galaxies, if the "gravitational constant" can be modified, the contradiction between the "dynamic mass" and "photometric mass" of galaxies can be well solved. This shows that the "gravitational constant" is not a constant, but a parameter reflecting the amount of "dark matter" in a galaxy.

Because "dark matter" not only does not have electromagnetic characteristics, but also exists in the form of a "field". For "dark matter", we cannot detect it directly, but can only obtain it indirectly through other methods. If the existence of the same physical quantity can be proved through different channels, it indicates that the physical quantity does exist.

Postscript

As a modern physical theory, this article has the following characteristics: first, it has improved the theory of light; second, it has put forward a new theory of space. Light is the medium for human beings to understand the material world. Space is the basis for the existence of the material world. Different interpretations of the physical properties of light and space will have a significant impact on existing physical theories.

The theory of light includes the principle of Doppler redshift of light, cosmological redshift, constant speed of light, and refraction, reflection, transmission and interference of light. These physical phenomena reflect the nature of light from different aspects. Each theory of light should at least be self- consistent. Any theory of light should meet the principle of "constant speed of light", "Doppler redshift" and "cosmological redshift" at the same time. The three principles together describe the course of light's "life". Moreover, the three principles are interrelated and mutually verified. The theory of light is correct only if it does not violate these three principles.

As the largest physical existence of the universe, space's physical characteristics will affect all physical behaviors. Is there an omnipresent and omnipotent "God" in nature? As a physical being, only space can have this qualification. Vacuum space regulates all physical behaviors in space and dominates all physical movements. The greatest contribution of this paper is to describe the physical properties of vacuum space from different physical perspectives.

1 If the motion reference object of the motion system cannot be correctly selected, the motion of the system cannot be accurately described. Modern physical theory discusses the motion of matter, so the motion reference frame is the basis of modern physical theory. The formation of modern physical theory began with the debate on Newton's absolute stationary reference system.

2 Einstein's principle of "the speed of light is constant" actually represents the view of Newton's absolute stationary motion refcrence system. Since the background space of the speed of light is electromagnetic space, taking the speed of light as the reference object of motion is to take "electromagnetic space" as the background space of motion. "Electromagnetic space" is a very important physical quantity,

without which motion cannot be defined. The speed of light must take "electromagnetic space" as the reference object of motion. Doppler redshift must also take "electromagnetic space" as the reference. Because the space is expanding, the "electromagnetic space" is also expanding synchronously. Only with the "electromagnetic space" as the reference object of motion, can we distinguish what is "relative motion" and what is "absolute motion".

3 What is "relative motion"? The motion satisfying the principle of relativity is "relative motion". "Relative motion" is the motion emphasized by mach. What is "absolute motion"? The motion relative to "electromagnetic space" is "absolute motion". Relative motion contains absolute motion. For example, the separation between galaxies. In the separation between galaxies, there are both "absolute motion" relative to "electromagnetic space" and the motion of synchronous expansion with "electromagnetic space". The motion that expands synchronously with "electromagnetic space" satisfies the principle of "relativity", but it is actually in a static state. Two kinds of motion with different properties represent different energy forms. It is very useful in astronomical research to distinguish two kinds of motions with different properties.

4 It is very important to fully understand the two different redshifts of light in astronomical observation. The universe is expanding and galaxies are separating from each other, which are all distinguished by the red shift of the spectrum. In the red shift of light, Doppler red shift and cosmological red shift are included. The Doppler redshift is far less than 1, and the others are cosmological redshifts. The Doppler redshift reflects the relative motion speed between galaxies, while the

cosmological redshift reflects the space distance.

5 Cosmological redshift and space expansion can be determined by the wavelength of the "cosmic microwave background" radiation wave. The visible light radiated during the Big Bang has become microwave. The growth ratio of electromagnetic wave length represents the rate of space expansion.

6 How to distinguish Doppler redshift in the red shift of light is a scientific research topic. If the Doppler redshift of light can be distinguished, the specific data of the gravity space a (t) diagram in Fig. 3.2 can be depicted.

7 It is incorrect to use Doppler effect to calculate Doppler redshift of light. Doppler effect is mainly to consider the relative motion velocity between the two, so as to conduct velocity transformation. It does not apply to the redshift of light. Because no matter how the light source or the observer moves, the speed of light between the light source and the observer will not change, which is called "constant speed of light". We can-not use "special relativity" to prove that "the speed of light is constant", or use the relativity effect to explain "the speed of light is constant". Because "constant speed of light" is the condition of "special relativity" hypothesis, Lorentz factor is derived from the basis of "constant speed of light".

8 The Michelson Morey experiment had a great influence on modern physics. It not only had an impact on Einstein, but also changed the research and development direction of later physical theories. The mistake of this experiment is that it does not take into account the blue shift of light. Is there a blue shift in the Michelson Morey experiment? This can be verified by experiments. According

to the principle introduced in this paper, in the Michelson Morey experiment, first adjust the orientation of the experimental device to form stable light interference stripes on the observation screen. At this time, both left and right rotations will cause interference fringes to gradually become disordered. The reason for the interference fringe disorder is that one of the two light paths has red shift and the other has blue shift. When the experimental device is rotated 180 ° , stable light interference fringes will appear again.

9 The principle of "constant speed of light" and Doppler redshift are two aspects of a mechanism. These two principles are combined to form the "relative motion effect". The "relative motion effect" indicates that if there is relative motion between the moving reference frame and the "electromagnetic space", the light will have a red shift. If the received light does not have a red shift, this relative motion will change the speed of light propagation in the reference system. "Relative motion effect" can be applied in many aspects, such as satellite networking, ranging, autonomous navigation and many other aspects.

10 Through the principle of cosmological red shift, we can get the conclusion that the "big bang" is the "big bang in space" as long as we invert the arrow of the time process. The "Big Bang" theory is more in line with the fact that space is expanding in reality. The theory can also make up for the inherent shortcomings of the original Big Bang theory, as well as the embarrassment of material being created out of nothing.

11 The Big Bang is an explosion of space, which undoubtedly proves that vacuum space is also a physical entity. Vacuum space is a physical entity that can explain many physical phenomena that cannot be explained before. With vacuum space as a physical entity, we can

well understand what is "dark energy" and why the universe expands.

12 Space is the largest subject in the universe. The biggest error in modern physics is to ignore its existence. There is no space connection in the universe, and all physical behaviors become isolated events. When we regard vacuum space as a physical entity, we can interpret many seemingly incredible physical phenomena. The key point of space described by Einstein's "general theory of relativity" is that it is a physical entity. The reason why matter has mass is that the space where matter exists is an entity. The reason why energy is conserved is also that space is a physical entity. Two particles thousands of miles away are entangled, and space is the link between them. Space contains everything and controls everything.

13 There is a misunderstanding in the understanding of the structure of matter in modern physical theory. Hundreds of so-called basic particles discovered by particle colliders are regarded as the basic structure of matter. In fact, these are material fragments. There are only two kinds of real basic structures of matter, one is fermion, which represents "the action of matter", and the other is boson, which represents "the energy of motion".

14 The theory of three forms of matter is the essence and soul of modern physical theory, which can well explain the composition and structure of the universe. The "equivalent equilibrium state" describes the organizational structure of the sun, moon and stars. "Non equilibrium state" explains the formation and nature of black holes. The "absolute equilibrium state" gives the physical properties of vacuum space. The visible universe consists of the sun, moon, stars, black holes and vacuum space. Although Newton's gravitational action

can be used to explain celestial bodies and ocean phenomena, Newton did not give the reason for gravitational action. The space theory of "absolute equilibrium state" can completely explain all problems of mass, gravity, inertia, etc.

15 The wave particle duality of microstructure lies in that it is a dual subject. When the subject of "material action" plays a role, the microstructure is a wave. For example, when an electron moves in an orbit outside the nucleus, it is the action of positive and negative charges. At this time, the electron is a wave. When light propagates in space, it is electromagnetic interaction. At this time, light is electromagnetic wave. "Material action" can only exist in the form of waves. When the energy subject plays a role, it is particles. For example, when an electron accepts the transition of energy from the ground state to the excited state, the energy works, and the transition of the electron is the property of a particle. When the electromagnetic wave of light encounters an obstacle, it will produce reflection or refraction. At this time, energy is acting, and light is expressed as a particle. Particles are characteristic of energy.

16 Finding a particle in an area is a probability event, but it does not represent the probability of the particle in the area. It is found that a particle is a non "elastic collision". Whether particles can be found in a space area is related to the distribution density of the particle's energy in the area. The probability of particle discovery is the probability of energy transfer. If the particle is not found, it does not mean that the particle does not exist.

17 The natural state of microstructure is the state under the action of "material action", which is a kind of wave. We cannot use the

physical parameters of a particle to describe the material wave in the natural state. The "uncertainty" principle actually uses the physical parameters of particles to describe the material wave, which brings "uncertainty" to the position and momentum. Whether its negative effect determines the causality in the objective world.

18 The natural state of microstructure is an "equivalent equilibrium state", which can collapse to "1" or "0". If the "equivalent equilibrium state" is considered to be a "superposition state", this "superposition state" has its specific physical connotation. There is nothing wrong with the principle of "state superposition", but some interpretations deny the logical relationship of the objective world. The most famous one is Schrodinger's cat.

19 Einstein spent the rest of his life trying to unify the four natural forces, which was caused by the misunderstanding of "weak force" and "strong force". Through the interpretation of the normative effect of vacuum space, we know that there are only two forces in the universe, universal gravitation and electromagnetic force. The weak force is the universal gravitation under the "vacuum field" specification, and the strong force is the electromagnetic force in the microstructure under the "vacuum field" specification.

20 It is generally believed that the mass energy formula expresses the mutual conversion of mass and energy, which is actually wrong. The mass energy formula belongs to the mathematical form in which the "vacuum field" regulates the universal gravitation in the microstructure. The mathematical expression of the normative effect of "vacuum field" on electromagnetic force in microstructure is Schrodinger equation.

21 Since microscopic "particles" are generally in potential wells, the role of potential wells is included in Schrodinger equation. If the potential well term in the equation is removed, the remaining equation term expresses the normative effect of the "vacuum field" on the electromagnetic force in the microstructure.

22 China has launched the Wukong "dark matter" detection satellite and established a "dark matter" laboratory at a depth of 2400 meters underground. All these may come to nothing. Because "dark matter" itself is only a "field". The seventh chapter of this paper describes the origin of "dark matter", the nature of "dark matter", the form of "dark matter", the energy of "dark matter", the evolution and detection of "dark matter". The above theory is based on the understanding of electromagnetic interaction. The point here is that electromagnetic force is the only force that enables matter to form a structure.

23 This is an article describing principles, involving many new physical concepts. Although some theories and principles are only a kind of inference, each inference can be proved from different aspects. The theory stated in this paper is not the ultimate theory, but a theory based on the current cognition and facts.

www.ingramcontent.com/pod-product-compliance
Lightning Source LLC
Chambersburg PA
CBHW011159220326
41597CB00026BA/4669